SEMA SIMAI Springer Series

Volume 36

As of 2013, the SIMAI Springer Series opens to SEMA in order to publish a joint series aiming to publish advanced textbooks, research-level monographs and collected works that focus on applications of mathematics to social and industrial problems, including biology, medicine, engineering, environment and finance. Mathematical and numerical modeling is playing a crucial role in the solution of the complex and interrelated problems faced nowadays not only by researchers operating in the field of basic sciences, but also in more directly applied and industrial sectors. This series is meant to host selected contributions focusing on the relevance of mathematics in real life applications and to provide useful reference material to students, academic and industrial researchers at an international level. Interdisciplinary contributions, showing a fruitful collaboration of mathematicians with researchers of other fields to address complex applications, are welcomed in this series.

THE SERIES IS INDEXED IN SCOPUS

Andreas Linninger • Kent-Andre Mardal •
Paolo Zunino
Editors

Quantitative Approaches to Microcirculation

Mathematical Models, Computational Methods and Data Analysis

Springer

Editors
Andreas Linninger
Departments of Biomedical Engineering and Neurosurgery
University of Illinois Chicago
Chicago, IL, USA

Paolo Zunino
MOX, Department of Mathematics
Politecnico di Milano
Milan, Italy

Kent-Andre Mardal
Numerical Analysis and Scientific Computing
Simula Research Laboratory
Oslo, Norway

Department of Mathematics
University of Oslo
Oslo, Norway

ISSN 2199-3041 ISSN 2199-305X (electronic)
SEMA SIMAI Springer Series
ISBN 978-3-031-58518-0 ISBN 978-3-031-58519-7 (eBook)
https://doi.org/10.1007/978-3-031-58519-7

This Springer imprint is published by the registered company Springer Nature Switzerland AG
The registered company address is: Gewerbestrasse 11, 6330 Cham, Switzerland

Preface

The microcirculation is a basic element of the cardiovascular system, comprising arterioles, capillaries, and venules with blood vessel diameters ranging from 5 to 100 μm. This intricate network is crucial for the transport and exchange of heat, respiratory gases, nutrients, waste, water, and hormones between blood and tissues. The dynamics of blood cells, including red blood cells, white blood cells, and platelets, govern energy supply to the organs and may hold the key several neurovascular diseases including stroke, hypertension, and dementias. Microcirculatory research is therefore a focal point in both physiology and medicine.

In microvessels, vital processes like the delivery of leukocytes and lymphocytes to tissue targets and the trafficking of inflammatory and immune cells occur. Microcirculation is also instrumental in regulating peripheral vascular resistance, vascular capacitance, and blood pressure, as well as controlling blood flow within the body's organs and tissues. Playing a pivotal role in maintaining bodily homeostasis and essential in oxygen transport to cells, the microcirculation is critical to overall health. Dysfunctions in this system can lead to various neurovascular diseases. Given its importance, microcirculation has been the subject of extensive theoretical and experimental research in recent decades, highlighting its central function in both health and disease.

The microvascular environment is a complex interplay of mechanical, hemodynamic, and regulatory elements, including capillary blood flow, blood kinetics within capillaries, capillary wall permeability, and diffusion phenomena in the tissue. Mathematical modeling plays a crucial role in quantifying and understanding the interactions among these factors, which are key to grasping both normal tissue function and the pathophysiology of various diseases. Most of the critical data about microcirculation come from in situ measurements in living animals or humans. To obtain a comprehensive functional understanding of microcirculation, a combination of different measurement techniques is essential. Mathematical modeling enhances the interpretation of these data, for example enabling the integration of in vivo measurements with ex vivo data on complete microvascular networks.

The complexity and multifaceted nature of microcirculation demand an interdisciplinary approach, encompassing medical imaging, continuum mechanics for modeling flow and tissues, applied mathematics, high-performance computing, and data analytics. A significant challenge in this field is to accurately represent phenomena occurring across multiple spatial scales.

Recent years have witnessed a major shift in how we approach biological systems modeling. Instead of simplifying their complex geometry, as traditionally done in engineering and mathematics, the focus has turned towards anatomically faithful mathematical models. These models aim to accurately represent the spatial, biophysical, and biological aspects of the natural system in intricate detail. This shift is driven by breakthroughs in medical imaging, which now enable us to observe vascular networks, blood flows, and transport phenomena with unprecedented resolution. This new era of anatomical modeling opens the door to virtually recreating entire organs on a computer. This development demands new capabilities from mathematical models. Two key challenges arise: (i) data integration: seamlessly merging medical image data with mechanistic models requires advanced techniques in image processing and machine learning; (ii) multi-physics and multiscale nature: anatomical models are inherently complex, dealing with multiple physical processes and operating across different scales. To tackle these requirements, mathematical modeling needs to go beyond simply increasing computational power. We need adaptive strategies and algorithms that can efficiently exploit the multifaceted structure of these problems. Multigrid and graph-based methods are expected to play a crucial role in this undertaking.

It is the focus of this collection to highlight the work of a select group of contributing labs who are making significant strides in addressing these challenges of spatially and functionally mixed-dimensional problems in the microcirculation. The collection of chapters within this volume highlights the cutting-edge trends in microcirculation modeling, particularly focusing on its significance in brain function analysis and the novel use of microvascular models in oncology. Readers will gain insights into the latest methodologies employed in microcirculation modeling, with a special emphasis on mixed-dimensional models. These models are particularly effective in delineating complex interactions in lower-dimensional spaces. Each chapter addresses the mathematical, computational, and practical aspects of these methods, demonstrating their ability to represent different aspects of microcirculation. Through this book, readers are offered a glimpse into the most recent advancements and techniques that are currently shaping the future of microcirculation studies.

The first three chapters of the book cohesively explore the application of mathematical models of microcirculation across diverse contexts: from cerebral blood flow and pathophysiology, through the nuanced dynamics of tumor microenvironments in radiotherapy, to innovative microvascular modeling techniques enhancing medical imaging and toxicity assessment.

More precisely, Chap. 1 delves into the cerebral microcirculation, emphasizing its complex network and significant role in nutrient delivery and waste removal in the brain. It explores the recent advances in understanding the cerebral microcirculation's anatomy and geometry, particularly the control of cerebral blood flow through pericytes, highlighting the role of these models in overcoming imaging challenges and suggesting future research directions. Chapter 2 focuses on the interplay of microcirculation and cancer treatment, in the context of radiotherapy. This chapter presents a computational model that examines the oxygen dependency and vascular network's role within the tumor microenvironment. It addresses the challenge of defining optimal radiotherapy schedules due to tumor heterogeneity. Chapter 3 introduces three in silico models based on a unified mathematical framework. These models simulate blood flow, drug transport, and their effects on cells in diseased microvascular networks. The chapter begins by deriving multi-phase multi-species flow and transport models from fundamental principles. It then applies these models to vascularized tissues, offering a methodology to connect changes in vascular network structures with microcirculation effects and macroscale biomarkers. These models are crucial for interpreting dynamic medical imaging and assessing drug toxicity, thereby bridging the gap between in vivo and in vitro experiments and reducing the need for invasive tests.

In Chaps. 4–8, the book addresses a range of advanced computational strategies. These include the application of finite element software for network model performance, the use of fast Fourier-type preconditioners that circumvent matrix inversion in massive transport problems across the blood brain barrier, and the deployment of novel numerical methods for microfluidics. Together, these chapters provide a thorough overview of advanced tools designed to address the computational complexities in microcirculation computational studies.

Initially, Chap. 4 assesses how finite element software, especially FEniCS and FEniCSx, manages mixed-domain partial differential equations for use in extensive hydraulic network models in microcirculation. Then Chaps. 5 and 6 address the important topic of numerical solution algorithms for the large-scale problems that arise in simulating large vascular networks. Chapter 5 presents a novel method for simulating oxygen exchange between cerebral blood and brain tissue. This problem has multi-physical character since solute transport in blood is convective, while in tissue it is governed by reaction and diffusion. Moreover, the vascular graph is thin, while the tissue expands over a three-dimensional domain. It introduces a new framework of Fast-Fourier Preconditioning in combination with a Schur Complement method to tackle coupling of solute transfer across multi-physics, mixed domains. It effectively models oxygen transfer from blood to brain tissue, integrating complex vascular network simulations with tissue metabolism and oxygen dissociation processes. Then, Chap. 6 develops a robust preconditioner for mixed-dimensional PDEs in 3D-1D domains using Lagrange multipliers, addressing the challenges posed by the coexistence of distinct length scales in micro-circulation and the complexity of the discretized system. Finally, Chaps. 7 and 8 address new computational techniques. Chapter 7 reviews and analyzes numerical models for multiphase flows in microcirculation, including the Volume of Fluid approach for fluid interactions and the dynamic Immersed Boundary method for bio-inspired systems, offering insights into multispecies mixtures and fluid/structure interactions. Chapter 8 investigates a numerical solution technique for an elliptic PDE with a Dirac source term, relevant to vascularized tissue flow models. The chapter explores the use of Zenger correction for regularization and proposes an iterative scheme for handling coupling terms in the equation.

Together, these chapters provide a comprehensive insight into advanced computational techniques and their applications in the simulation and analysis of complex microcirculatory systems.

Chicago, IL, USA — Andreas Linninger
Oslo, Norway — Kent-Andre Mardal
Milan, Italy — Paolo Zunino

Contents

Mathematical Models of the Cerebral Microcirculation in Health and Pathophysiology

Stephen J. Payne

Abstract The cerebral microcirculation is a highly complex interconnected network of vessels that cover a wide range of length scales and respond over a wide range of time scales. It plays a critical role in the delivery of nutrients to tissue and the removal of metabolic waste products. Impairments to the cerebral microcirculation are implicated in many cerebrovascular and neurodegenerative diseases. Despite this, until around 15 years ago, little quantitative information was known about its anatomy and geometry in humans; the advent of new data since then has led to a substantial body of work to model blood flow and oxygen transport within the microcirculation and surrounding tissue. Until around 10 years ago, it was also thought to be a largely passive bed with control of cerebral blood flow (CBF) occurring upstream (in the arteriolar bed) and with limited implications in the impairment of cerebral blood flow, either in healthy or in diseased states. However, it is now known that there is significant control of CBF at this length scale through the action of pericytes. Given the difficulties involved in imaging this vascular bed in humans, models of the cerebral microcirculation have an important role to play in gaining a better understanding of CBF and its behaviour in different conditions. In this chapter, the development and application of models of the cerebral microcirculation are thus described, together with potential avenues for future exploration.

Keywords Cerebral blood flow · Cerebral haemodynamics · Microcirculation

1 Introduction

The cerebral microcirculation covers a wide range of different generations and hence length scales, from vessel diameters of a few hundred micrometres to under 10 μm. After arterial blood enters individual territories via the leptomeningeal

S. J. Payne (✉)
National Taiwan University, Taipei, Taiwan
e-mail: stephenpayne@ntu.edu.tw

A. Linninger et al. (eds.), *Quantitative Approaches to Microcirculation*,
SEMA SIMAI Springer Series 36, https://doi.org/10.1007/978-3-031-58519-7_1

vessels on the cortical surface, penetrating arterioles dive into the cortical tissue. These vessels bifurcate, monotonically decreasing in diameter with accompanying changes in wall composition, until the capillary bed is reached. The capillary bed is made up a highly interconnected network of vessels that dynamically adapt to ensure that no brain cell is more than approximately 25 μm from a capillary vessel [1].

Blood leaving the capillary bed then passes through a reverse process with vessels joining together until blood reaches the ascending venules and hence passes back into the venous macrocirculation. The penetrating arterioles and ascending venules are found in an approximate 3:1 ratio in humans [66] and each form specific territories that are supplied by individual units, thus giving a fundamental length scale for flow in the microcirculation. A schematic of the topology of the different generations is shown in Fig. 1 and a schematic of the configuration of the different generations in Fig. 2.

The cerebral cortex is the surface layer of brain tissue, and although only 2–4 mm thick, this layer plays a key role in multiple brain processes (including memory, perception, and language). The layer is highly folded to provide a very large surface area and is comprised of four lobes (temporal, occipital, parietal, frontal). The surface layer of the brain is termed grey matter and the deeper material is termed white matter. The names reflect the shading shown in imaging data and the different compositions: grey matter has few axons but many neuronal cell bodies, with white matter the reverse.

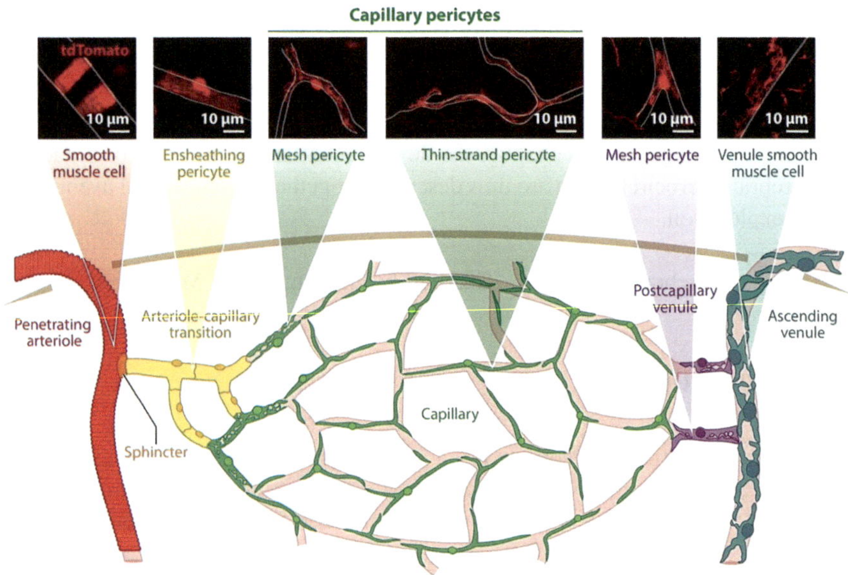

Fig. 1 Schematic of different microvascular generations in the cerebral circulation, reproduced with permission from Hartmann [32]

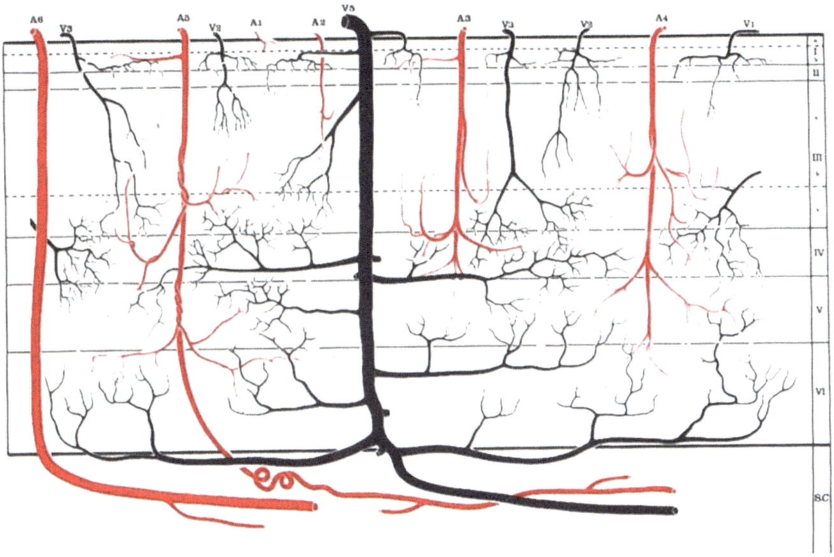

Fig. 2 Typical geometric arrangement of microvascular generations in the cerebral cortex, reproduced with permission from Duvernoy [16]

Different vessels supply and drain different cortical layers, and some vessels pass through the entire cortical layer before supplying the capillary bed; this being necessary to provide sufficient blood to the deeper tissue. The capillary vessels comprise both an endothelial layer (where endothelial cells that are coupled by tight junctions form the blood-brain-barrier) and the basal membrane. As shown in Fig. 1, the walls of the larger blood vessels in the microcirculation also contain smooth muscle cells that act to regulate vessel tone and hence diameter. The smaller blood vessels are controlled via different types of pericyte, the role of which in controlling blood flow has recently become much clearer [5, 31]. The whole circulation is a highly dynamic, active system that responds to multiple global and local stimuli across a very wide range of time scales (from seconds to years) to maintain a tightly coupled relationship between tissue oxygen supply and demand.

Mathematical models of the cerebral microcirculation were first developed in their current form around 2006. This was prompted by the publication of the first quantitative experimental data on the (ex vivo) human cerebral microcirculation by Cassot [12], in parallel with the development of new imaging modalities that enabled experimental data to be gained (in vivo) in small animal models. Techniques that have subsequently been applied in this context include two-photon microscopy with laser ablation to image fluorescently labelled microvasculatures [75, 76], synchrotron radiation-based X-ray microscopy to obtain tomographic microvascular images [30, 62], micro-optical sectioning tomography [85], selective plane illumination microscopy [21], optical coherence tomography [47], and three-photon microscopy [38].

An important point to note in these measurements is the difference between the anatomical and geometrical properties of the human and non-human cerebral microcirculations. A detailed review of these experimental data has been provided by Schmid [66], illustrating that care needs to be taken when comparing results from human and non-human studies. There have been no detailed studies that compare the flow fields between animal and human, although scaling relationships have been used to assist comparison between human-based simulations and animal experimental data [86].

Models of the cerebral microcirculation can be roughly divided into essentially three types: the first that uses experimentally measured networks; the second that uses synthetic networks based on experimental data; and the third that uses image-based circulatory network synthesis (iCNS) methods that are a combination of imaging information and perfusion and haemodynamic construction principles. The first approach has the advantage of being based directly on real networks but is limited to very small volumes (which means that the choice of boundary conditions can strongly influence the flow fields) and is restricted to animal models. The second approach has the advantage of being able to study flow fields in networks of arbitrary size but is limited by the assumptions made to generate such synthetic networks (although these can be in turn based on human data). The third method aims to combine the advantages of both experimental data and general principles in network construction, although care must be taken to combine these in a rigorous way, and these are currently limited by being based on animal data. One example of this approach was proposed by [34], where 100 synthetic networks were created that matched the most relevant properties of the experimental data from 14 different samples from the mouse cortex.

Approaches to generate synthetic networks then also tend to fall into one of two categories. The first approach grows networks from simple seeding distribution, based on a set of heuristic rules, see for example [2], whereas the second approach generates networks that aim to satisfy given statistical distributions for the network properties, see for example [17]. These two approaches are essentially very similar, although the latter tend to focus more on the optimisation process and can be highly complex, whereas the former generally aim to provide insight into the mechanisms by which the networks are created.

All these approaches are limited by the availability of experimental data, in particular human experimental data, although more recently there has been some work to make networks publicly available (including the 14 mouse cortex networks and 100 synthetic networks described above), which opens up the possibility of further data sharing and comparative studies in future. Extracting accurate networks from experimental data is a highly challenging task and small errors in the geometry of these networks can result in large errors in flow field (due to the fourth power dependence of vessel conductance on radius). Segmentation at this length scale involves dealing with dense networks with complex connectivity, often compromised by poor signal-to-noise ratios, movement artefacts, and a limited depth penetration. As a result, significant amounts of manual post-processing are often required [24].

The first study that provided quantitative data for the human microcirculation was that of Cassot [12] (a similar study in the macaque was performed at a similar time by Weber [84]). The authors re-analysed the experimental data originally acquired by Duvernoy [16], who used thick sections of human brain injected with Indian ink and confocal laser microscopy. Through the acquisition of hundreds of thousands of segments, Cassot and colleagues calculated the distributions of vessel diameter and length, the vessel density and vascular volume. Other important parameters include the vessel tortuosity, orientation, diameter ratios and branching angles. The networks predominantly comprise bifurcations, with relatively few (~5%) trifurcations.

The resulting networks obtained were subsequently used by Lorthois [45, 46] to explore the concept of vascular territories, in terms of the regions of the networks that are supplied by the largest vessels in the microcirculation (i.e., the penetrating vessels). The role of the venous circulation in the BOLD-fMRI response and the relationship between pial draining vein diameter and the volume of a neuronally activated area were also shown, and models of the activation response have subsequently been constructed from anatomically accurate networks from first principles [23]. Figure 3b shows a reconstructed network based on imaging data, Fig. 3a, that has been used to compute both blood flow and oxygen transport, Fig. 3c–e.

Various systems have been proposed to label vessel generations, for example the Strahler system and the Kassab system, which can then be used to propose scaling relationships for vessel properties using results such as those shown in Fig. 3f. Based on these relationships, microcirculatory models can be constructed, for example using Voronoi tessellation [64, 71], constrained constructive optimisation [43], minimum spanning trees, [18, 72], angiogenesis-based methods [67] and double bifurcation closures [34]. A (partial) list of models of the cerebral microvasculature is given in Table 1 showing how these models have gradually increased in size and accuracy. In the next section the methodologies used to calculate the flow fields will be described.

One potential avenue for exploration in the future is the use of deep learning to improve segmentation, building on approaches in other microcirculation systems, see for example [73]. More recent imaging studies have opened the possibility of imaging whole organs using HiP-CT [81], but there are no quantitative data available from such studies yet. Such future data sets will also provide more information about the variations in anatomical properties across different brain regions; it is thought that the properties vary little over short distances but may well be very different in different brain regions, see for example the data of Shaw [70].

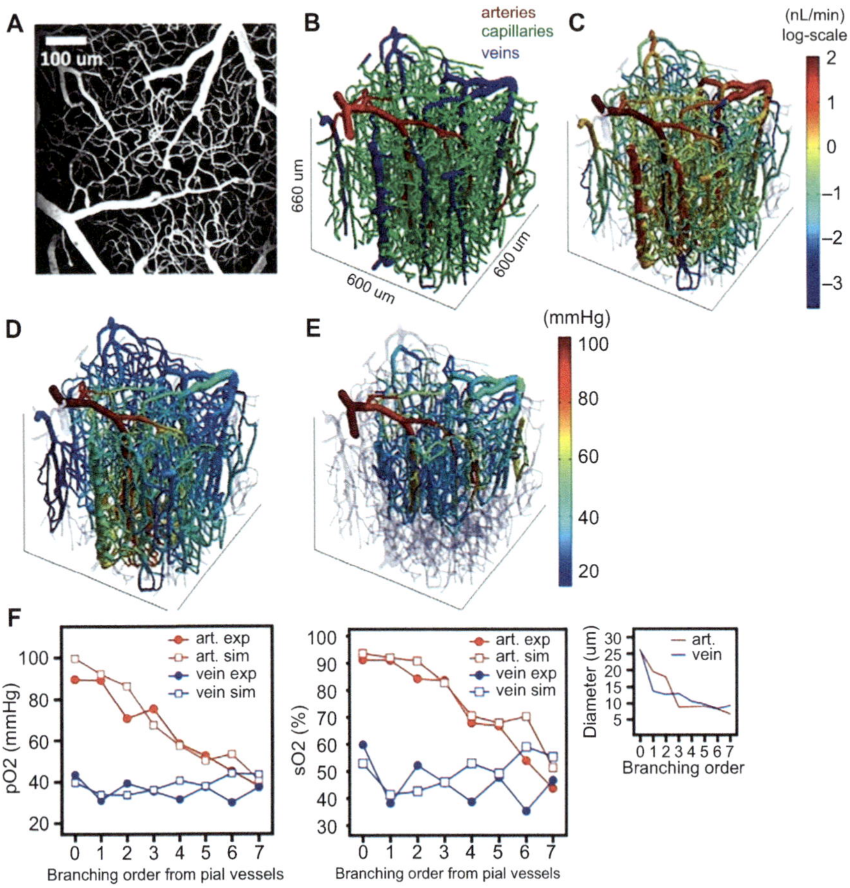

Fig. 3 Construction of realistic vascular networks. (**a**) TPM FITC angiogram of the mouse cortex. (**b**) FEM mesh of the vasculature displaying arteries, capillaries, and veins. (**c**) Blood flow distribution simulated across the vascular network assuming a global perfusion value of 100 ml/min/100 g. (**d**) Distribution of the partial pressure of oxygen (pO2) simulated across the vascular network using the finite element method model. (**e**) TPM experimental measurements of pO2 in vivo using PtP-C343 dye. (**f**) Quantitative comparison of simulated and experimental pO2 and SO2 distributions across the vascular network for a single animal. Traces represent arterioles and capillaries (red) and venules and capillaries (blue) as a function of the branching order from pial arterioles and venules, respectively. Figure and title reproduced with permission from Gagnon [23]

Table 1 List of models of the cerebral microvasculature. N.B. the type of model refers to the architecture/geometry of the model. Adapted from Payne [55]. Model capability: 1 = flow-pressure simulation; 2 = variable haematocrit and plasma skimming; 3 = prediction of blood and tissue oxygen

Author	Species	Capillaries	Penetrating vessels	Model type	Number of vessels	Model size—length × width × depth (μm)^3	Model capability
Zagzoule and Marc-Vergnes (1986)	Human	Y	Y	Lumped parameter, 2D	316,858	Whole brain	1
Secomb et al. (2000)	Rat	Y	N	Cast, 3D	50	140 × 150 × 160	1, 3
Beard (2001)	Rat	Y	N	Cast, 3D	50	150 × 160 × 140	1, 3
Boas et al. (2008)	Rat	Y	Y	Vascular anatomical model, 2D	254	N/A	1, 2, 3
Fang et al. (2008)	Rat	Y	Y	Cast, 3D	N/A	230 × 230 × 450	1, 3
Reichold et al. [62]	Rat	Y	Y	Cast, 3D	N/A	N/A (volume 2.8 mm^3)	1, 2, 3
Guibert et al. [30]	Marmoset	Y	Y	Cast, 3D	16,000	N/A (volume 18 mm^3)	1, 2
Lorthois et al. [45]	Human	Y	Y	Cast, 3D	10,318	N/A (volume 1.6 mm^3)	1, 2
Su et al. (2011)	Human	Y	N	Statistical, 3D	1073	500 × 500 × 500	1, 2
Safaeian and David [64]	Human	Y	N	Statistical, 3D	492	600 × 380 × 250	1, 2, 3
Blinder et al. (2013)	Mouse	Y	Y	Cast, 3D	25,498 (average)	1200 × 1450 × 1900	1
Linninger et al. [43]	Human	Y	Y	Both, 3D	256,000	3000 × 3000 × 3000	1, 2, 3
Gagnon et al. [23]	Mouse	Y	Y	Cast, 3D	N/A	600 × 600 × 662	1
El-Bouri and Payne [17]	Human	Y	N	Statistical, 3D	2000	625 × 625 × 625	1
El-Bouri and Payne [18]	Human	N	Y	Statistical, 3D	1000+	1000 × 1000 × 2500	1
Gould et al. (2017)	Mouse	Y	Y	Cast, 3D	23,496 (average)	1130 × 1130 × 1480 (average)	1, 2, 3
Schmid et al. [65]	Mouse	Y	Y	Cast, 3D	23,496 (average)	1130 × 1130 × 1480 (average)	1, 2
Hartung et al. [34]	Mouse	Y	Y	Statistical, 3D	Whole brain	Whole brain	1, 2, 3

2 Methodology

Once the networks have been constructed, the key parameter that contains the anatomical information at this length scale is the permeability tensor. This comes from the homogenisation of stakes flow in vessel networks in terms of Darcy flow, i.e.:

$$\mathbf{u} = -\mathbf{K}\nabla p \tag{1}$$

where the flow velocity, $\mathbf{u}$, is proportional to the pressure gradient, ∇p, via the permeability tensor, $\mathbf{K}$. As the permeability tensor is independent of the boundary conditions applied, it is also a convenient parameter to compare between different networks and can easily be computed for any arbitrary network of any size. Changes in this parameter have also been quantified in different conditions, as described below. A more detailed derivation of this can be found in [17, 54].

There has been some work to compute the transit time distributions in the microcirculation and to explore how these are affected under different conditions. Park [53] proposed a theoretical framework for calculating both transit time distribution and oxygen extraction fraction, building on the work of Østergaard and colleagues, see for example [4, 39]. Goirand [26] more recently proposed a power-law-scaling approach, based on simulations in microvascular networks, which was used to compute transit time distributions and hence to describe network-driven anomalous transport and the emergence of critical regions (which could either lead to hypoxia or to the build-up of proteins such as amyloid-beta), Fig. 4. As transit time distributions can be directly calculated from residue functions, obtained using imaging data, this is an additional way of gaining information about the microcirculation that remains relatively under-explored to date.

To compute the flow, several approaches have been adopted, essentially either assuming a continuum approach or a discrete approach. In the continuum approach, empirical correlations are assumed for viscosity as a function of vessel diameter and haematocrit, for example [60, 61]. Haematocrit splitting relationships at bifurcations are then also required, for example those proposed by [27, 61]. Conservation of flow and red blood cells at nodes, together with continuity of pressure, is then used to compute the flow field using an iterative approach, although many studies assume a constant haematocrit, which means that the flow field can be directly computed.

However, it has been shown in several modelling studies that haematocrit is spatially quite heterogeneous, varying with depth, see for example [26, 33]. Although this has been shown to have only a relatively small effect on the capillary permeability (unpublished results), the fundamentally short length scales over which oxygen transport and metabolism occur mean that fluctuations at very small scales can potentially have a very important role in the hypoxia-ischaemia response [86]. Whether or not the capillary bed regulates to maintain approximately constant haematocrit as opposed to constant flow remains an open question.

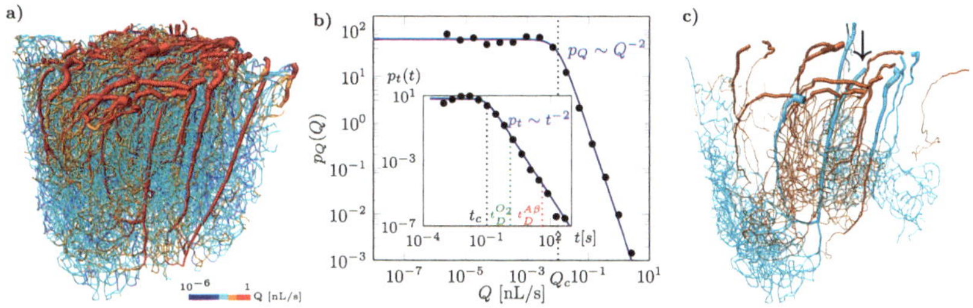

Fig. 4 The blood flow organization in cortical microvessel networks induces broadly distributed flow rates and vessel transit times. (**a**) Simulated blood flow map in a microvascular network feeding 1 mm^3 of the mouse cortex. Vessel flow rates are represented with blue shades for $Q < Q_C$ and red shades for $Q > Q_C$. (**b**) PDF of simulated flow rates (black dots) compared with the approximation (continuous blue line). The characteristic flow rate Q_C separating the uniform and power-law regimes are indicated as a dotted line. Inset: PDF of vessel transit times. The diffusion coefficients for oxygen and amyloid-β ($D_{O_2} = 2 \times 10^{-9}$ m^2 s^{-1} and $D_{A\beta} = 6 \times 10^{-11}$ m^2 s^{-1}) yield diffusion times ($t^{O_2}{}_D$ and $t^{A\beta}{}_D$) indicated by the green and red dashed lines, respectively. (**c**) Example of trajectories visiting <30 vessels (orange) and >70 vessels (blue), originating from the arteriole shown by the arrow. Figure reproduced and title adapted with permission from Goirand [26]

Limitations to this approach include the approximate nature of the empirical relationships, which are based on limited data samples, and the use of the Poiseuille equation, which assumes no vessel tortuosity. It is known that tortuosity is altered in many conditions, for example being higher in stroke patients compared to age-matched healthy controls [15]. The presence of tortuosity can also have a very significant effect on the resistance to flow at this length scale [51].

The empirical correlations that are commonly used for modelling purposes have a number of limitations, which make improving the accuracy of the relationships of great importance. These (highly non-linear) empirical relationships are based on data obtained from experiments in the rat mesentery (with the experimental data points not having yet been made publicly available) and their applicability to the cerebral microcirculation remains untested. Quantification of the errors in flow field caused by errors in these empirical relationships (which have been published in several different forms) has yet to be performed and their inclusion in network models appears to cause numerical convergence issues, as reported by a number of authors.

Discrete simulations have until recently been limited by the very substantial computational power required for even relatively small networks. However, simulations that couple the mechanical properties of red blood cells, white blood cells and platelets with the fluid dynamics of the plasma flow have recently started to become feasible. Models such as Hemocell [79, 88] have exploited the combined lattice Boltzmann and immersed boundary method methodology proposed by Kruger [42], and Hashemi [36], amongst others, to simulate the transport of individual blood cells through simple capillary networks. These simulations have the potential to provide a much more detailed understanding of the flow at this length scale. Other studies have adopted a simpler approach, using a rigid particle approach to individual cells through more complex networks, for example [33, 65] showing that the pathways through such networks are functions of the haemodynamic relationships and that there are thus potentially preferred flow pathways [44].

The use of a porous medium approach for the capillary bed based on a homogenisation approach, enables simulations of perfusion to be performed easily at length scales comparable to imaging voxels (i.e., of order 1 mm). However, this introduces a coupling between a discrete flow model and a continuous flow model, which is a non-trivial consideration. Peyrounette [57] proposed one solution to this, using analytical approximations of the pressure field to mimic the rapidly varying pressure fields around the coupling points, showing good accuracy at larger scales at a computationally reasonable cost. Other solutions include the use of a constant pressure boundary condition between the two domains [14, 41], although this is likely to be a limiting factor in simulations of pathological conditions.

The major assumptions made with the homogenisation approach are first that the behaviour is periodic at the local scale, which is clearly an approximation, and second that the behaviour can be locally averaged at a length scale somewhere between the local and global scale. Some of the detailed information will thus be lost (or 'smeared out') via this approach and it is not yet clear what effect this has on the behaviour at the different length scales. Detailed comparisons between the results obtained using network simulations and those obtained using the porous medium approximation will be needed to enable the accuracy of this approximation to be quantified.

An approach based on correlation length for a distribution of sources/sinks has recently been proposed that uses ensemble averaging to provide a homogenization approximation for the ensemble mean concentration [59]. This offers a method to include the spatial information about the distribution of sources and sinks using a correction term to the governing equation. Although this is currently only used for oxygen delivery, it could also be applied to flow calculations to provide a correction term to the permeability tensor.

Once the flow field has been computed, the transport of oxygen to tissue can be calculated using a mass-transport approach. Multiple techniques have been used for such calculations. Early studies used a Green's function approach, as this has the advantage of not requiring a discretization of the space; however, it requires the governing equations to be linear and rapidly becomes computationally intractable for larger volumes, as well as generally only being used in steady state due to the computational expense of dynamic simulations in comparison with other techniques. Finite element/finite volume approaches have been used by some authors, for example [24], although these have yet to satisfy the necessary convergence criteria for multiscale resolution.

More recently there has been a substantial amount of work performed on the family of dual mesh techniques for tetrahedral meshes. These techniques were recently combined with fast Fourier preconditioning to enable the prediction of both intravascular and extravascular partial pressure of oxygen at the micrometre length

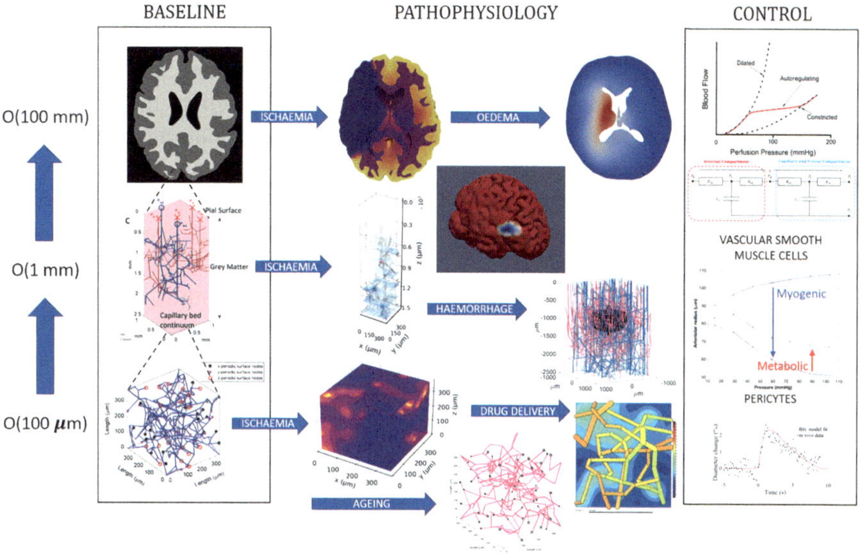

Fig. 5 Schematic of modelling of cerebral microcirculation and its role in baseline conditions (left) and in pathophysiology (middle), together with the control of cerebral blood flow (right), across different length scales (increasing from bottom to top)

scale, see for example [35] and [80]. These techniques open up the possibility of more detailed 1D-3D coupling than has been possible up to now.

In the remainder of this chapter, the focus will be on the use of statistical models. This approach enables the generation of a Representative Elemental Volume (REV) and hence the computation of the permeability tensor of the capillary bed through homogenization techniques. Importantly, this then enables the microcirculation behaviour to be applied at larger length scales within whole brain models [40, 41, 52], and hence changes in the geometrical and anatomical properties at this length scale to be included within simulations at larger length scales, as shown in Fig. 5. The statistical approach has also been extended to the penetrating vessels [14, 18, 19], which enables both the passive and active behaviour of the microcirculation to be considered, again within the full range of length scales in the cerebral circulation in a computationally feasible way. Such an approach enables the microcirculation to be fully integrated within whole brain models and hence to be more easily and more directly related to experimental data.

3 Applications

In this section, the numerous applications of statistical models of the cerebral microcirculation are illustrated and potential future avenues for exploration are described. A schematic of this approach is shown in Fig. 5, illustrating how these models play an integral role in models of the wider cerebral circulation, both in healthy conditions and in response to pathophysiological challenges.

Statistical models have recently been used to quantify the impact of microthrombi (which can be formed following mechanical thrombectomy after acute ischaemic stroke) on cerebral perfusion [20]. This shows that the type of thrombectomy technique and stiffness of the clot both have a strong influence on the drop in cerebral perfusion. These simulations also highlight that the response shows a high degree of variability, given the randomness in both the anatomy of the microcirculation and the size and number of the microthrombi, indicating one important reason why there is such significant variability in the clinical response. In addition, the response shows a strong depth-dependence, which mirrors other studies that show strong depth-dependence on haematocrit distribution [44].

These results show that micro-emboli can play an important role in the no-reperfusion phenomenon following successful removal of a large clot, although there will be other factors, such as inflammation and pericyte ischaemia-induced constriction, that will likely also be involved. Future work will be needed to quantify these effects to understand the relative contributions to the no-reperfusion phenomenon under different conditions, opening up the possibility of designing interventions to inhibit this phenomenon.

A Green's function approach was then applied to quantify the effects of cerebral microthrombi on oxygen transport to tissue, to quantify the hypoxic response to ischaemia [86], Fig. 6. Although the drop in perfusion is a linear function of the proportion of blocked vessels, with a scaling factor of around 3%/% for small changes, the tissue hypoxic fraction is a strongly non-linear function of blockage fraction with minimal hypoxia (defined as the fraction of tissue with an oxygen partial pressure below 10 mmHg) below a blockage fraction of 10%. However, this increases rapidly thereafter, showing that the capillary bed has a certain degree of

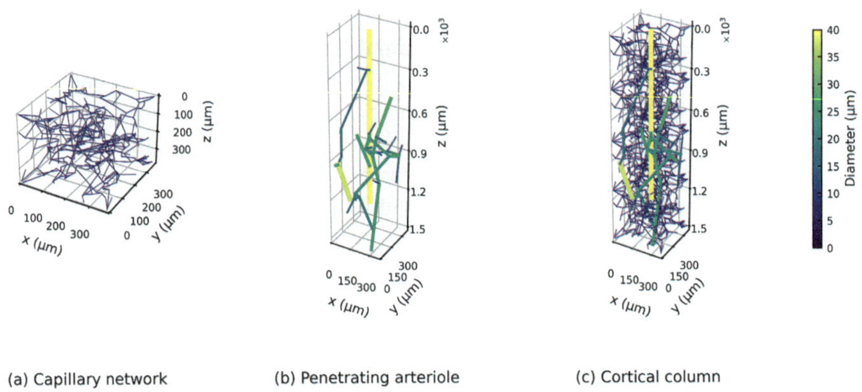

Fig. 6 Geometry of a typical cortical column (**a**) and its tissue oxygenation (**b**, **c**) solved by the Green's function method. Reproduced with permission from Xue [86]

redundancy, and that perfusion can reduce by around 30% due to microthrombi before hypoxia starts to become evident and by around 50% before the hypoxic fraction reaches 50%. In these simulations, however, since no active response is included, these are likely to be the 'worse case' levels of redundancy, and more experimental data will be required to quantify the effects of regulation at this length scale.

Comparisons with results from small animal models show excellent agreement, indicating that the behaviour at this scale is comparable between different species [86]. In these experiments, microspheres (diameter 25 μm) were injected through the carotid artery to occlude penetrating arterioles. Segmentation of the locations of both the microspheres and the resulting hypoxic regions showed that the relationship between the two is more complex than originally hypothesised, with significant distances occurring between the two. Numerical simulations showed the same characteristic type of behaviour when distance analyses were performed and these were found to be mainly due to the hypoperfusion that occurs in the region downstream of the occlusion site.

One other important physiological factor that governs this response is the extravasation of the microspheres into the extravascular space; this has been shown by other experimental studies, see for example [78], to have a time constant of a few days. Although this means that extravasation is unlikely to play an important role in the acute phase of ischaemia, it could be an important process in the longer-term response. However it is not yet known whether this phenomenon is altered in different conditions and how this relates to the inflammatory response, which has yet to be modelled in any detail in this context.

To investigate the dynamic changes in the permeability of the microcirculation, statistical networks have been applied to data from pig models of the cardiac microvasculature dynamic response to myocardial infarction (MI) to calculate the time- and region-dependent alterations in tissue perfusion [25]. There is evidence of a rebound in permeability at 7 days post MI compared to 3 days post MI (although both values are below 1 day post MI), indicating that there are processes that act to restore the health of the microvasculature (potentially including extravasation).

One of the most common complications of ischaemic stroke is haemorrhagic transformation (HT); this is caused by dysfunction of the blood-brain-barrier and can be exacerbated by thrombolytic therapy [3]. It is also thought potentially to be implicated in local rises in intracranial pressure (ICP) and to result in capillary compression, which may be part of the no-reflow phenomenon [10]. Using a Green's function approach, simulations of HT in statistical models of the cerebral microvasculature have shown that a volume of haematoma forms that is roughly proportional in volume to the volume of the affected vessel [82], and that this does appear to act to compress the microcirculation with a resulting drop in both perfusion and CBV [83]. Most recently, this model of HT has been incorporated within a whole brain model to enable comparison to be made to clinical imaging data, again highlighting the value of the multi-scale approach in linking the behaviour at the microcirculation length scale to the whole brain length scale (manuscript in press).

Another component of the response to ischaemic stroke is oedema. This comes in two forms, vasogenic and cytotoxic, and these both dynamically change post ischaemia, requiring careful management of patients [69]. Although parts of this response appear to be intrinsically harmless, other aspects of the response can be very damaging to brain tissue. The relationship between increases in intracranial pressure and flow permeability of brain tissue (due to vessel compression) has been investigated in a couple of preliminary studies, see for example [50]; however, more detailed studies of the relationship will be needed to gain a fuller understanding of the oedemic response and its role in the multiple mechanisms that govern the dynamic response to ischaemia.

There have been many models developed of cerebral autoregulation (the mechanism that acts to maintain cerebral blood flow near constant in response to short-term changes in arterial blood pressure), see for example [6, 74, 77]. These have nearly all only considered the regulatory mechanisms that act to adjust vascular tone in the arteriolar bed, although it is now well known that pericytes play a key role in the regulation of perfusion within the cerebral microvasculature. A better understanding of the interaction of these two responses (since it now seems clear that both have a role to play, but their relative importance is not yet clear) will be important in interpreting the dynamic response to both global stimuli such as arterial blood pressure and blood gas levels, and local stimuli such as the neurovascular response. The first such combined model has recently been proposed, based on the response of small animal models to visual stimulation [48], showing that capillaries do appear to play a fundamental role in the rapid response to changes in blood pressure, particularly in ensuring that the response time is sufficiently rapid (since the time constant for dynamic cerebral autoregulation is around 5–10 s).

More accurate mathematical models of the pericyte response and its effects on capillary permeability will, however, be needed so that they can be incorporated within existing whole brain models and thus compared to time series data for dynamic cerebral autoregulation. Such an approach will provide insight into the regulation of blood flow across the whole cerebral vasculature, building on existing (well-validated) models that focus on the arteriolar vessels, see for example [6, 74]. However, such models are highly complex, and the pathways remain not fully understood, indicating that both animal models and human studies will be required fully to disentangle the regulation response. The importance of spatial transmission of signals through the vasculature (i.e., the co-ordination between different vessel generations) will need to be better understood, as will the role of the sympathetic nervous system in the co-ordination of the response: see [54] for a recent summary of these different mechanisms.

The role of pericytes in the ischaemic response has been measured experimentally, with pericytes acting to constrict flow in response to hypoxia-ischaemia [31]. The non-linear response between vessel blockage and permeability discussed above means that this may have a significant effect, although potentially only in severe or prolonged hypoxia. Future work includes modelling the response of pericytes to ischaemia and again quantifying the effects of this on permeability and hence on perfusion in a whole brain model. The permanent occlusion of microvessels through pericyte constriction has been suggested as one part of the response to both stroke and dementia [31], although since these take place over very different time scales, it is likely that the mechanisms will be different.

The effects of ageing on the cerebral circulation have recently been measured in mice, providing valuable information about the ageing microcirculation. Reductions in microvascular density [11], plasticity [63], haematocrit, and capillary density have all been shown, alongside increases in capillary transit time heterogeneity [49]. Through a choice of scaling rules and conversion from mouse to human age, the effects of ageing have been included in one recent model of the microcirculation [29]. Blood vessel surface-area-to-volume ratio is found to reduce in old age by 6% and permeability by 24% from middleage to old age. The spatial variability within the microcirculation was also found to increase with age, with a threshold being found around 75 years old, after which small changes have an amplified effect on perfusion parameters.

These results point to a vulnerability to changes that is similar to the robustness to microthrombi (i.e., that the system is able to cope with changes up to a certain threshold, beyond which 'tipping point' small changes have exaggerated effects), which needs to be carefully considered in models of the cerebral microcirculation. This is particularly important given the fact that most cerebral diseases have a significantly greater incidence with increased age and thus studying the response to diseases such as stroke must consider the age and general health of the patient, particularly in terms of the robustness to stimuli.

In the context of therapy, modelling drug delivery at this length scale provides a valuable in silico approach that can be used to optimize therapy before carrying out in vivo (or indeed animal) studies. Blood-mediated nanoparticle delivery has the potential to facilitate crossing the blood-brain barrier. A modelling approach, building on the earlier results, can be used to include clearance and cellular uptake of a chemotherapeutic to assess optimal infusion parameters, with experimental validation in progress, and to assess the potential of (for example) gold nanoparticles to cross the blood-brain-barrier, noting that the BBB is itself dynamically changing post ischaemia [13].

4 Future Directions

The relative paucity of data for the anatomy and geometry of the microcirculation has already been mentioned, and this remains the largest source of uncertainty in models of the cerebral microcirculation, although data sharing has now started to occur in this context. The advent of new imaging modalities that can map entire cerebral circulations ex vivo has opened the possibility of much larger data sets than have been available in the past, but there is still a need to obtain more detailed information about the brain in vivo. In this context, however, the use of parameters such as transit time distribution does provide potential for additional information to be obtained and used clinically.

The active behaviour of the cerebral microcirculation has only recently started to be appreciated and this also opens new possibilities for a better understanding of how this network acts in response to multiple stimuli and how this takes place over multiple time scales (from seconds in the context of autoregulation to decades in the context of neurodegenerative diseases). New mathematical tools will need to be applied to understand better the processes that occur over such different time scales.

The lack of certainty over some of the key physiological parameters, such as the diffusion coefficient of oxygen in tissue and the dependence of oxygen metabolism on partial pressure of oxygen, both of which play an important role in the supply of oxygen to tissue, remains a limitation to modelling oxygen transport to tissue. These parameters are, however, challenging to measure experimentally. One potential avenue for exploration is the consideration of length scales of order of magnitude the cell characteristic length (i.e., of order 10 μm). At this length scale, it is known that the movement of oxygen molecules is highly anisotropic, see for example the recent review by Pias [58]. In particular, the diffusion coefficient is known to be very different in different components of the cell.

Molecular simulations can be performed at this length scale to study the movement of oxygen molecules, opening up the possibility of calculating parameter values using a homogenisation approach across three length scales. Similar studies have been performed to investigate homogenisation of the mechanical properties of tissue, see Penta [56], and such techniques could be applied to oxygen diffusion. Note that it is well-known that a material can be isotropic at the microscale and anisotropic at the macroscale [37]. Such an approach could also be considered for the movement of substances other than oxygen and could help to explain the behaviour found at the 100 μm length scale.

5 Conclusions

In this chapter, the cerebral microcirculation has been considered in detail, examining its structure and its function, as well as how blood flows through it and how this is controlled and affected in different physiological and pathophysiological conditions. The recent development of mathematical models of the microcirculation and the use of multi-scale techniques has led to the development of blood flow simulations within the human brain across multiple length scales at reasonable computational cost. These models have started to be used to simulate multiple different pathological conditions: although we have focused on ischaemic stroke here, the framework can in future be used for a wide variety of other diseases since changes in the microvasculature can be included within this framework in a principled manner. Future work should consider how the models proposed here can be exploited over these multiple time scales (moving from seconds/minutes to years/decades) and hence to explore a much wider range of pathological conditions. Extensive further validation will however be required, particularly to understand the effects of different microvascular properties in different regions of the brain, and to better understand the different mechanisms that control cerebral blood flow at different length scales.

Acknowledgments SJP is supported by a Yushan Scholarship from the Ministry of Education, Taiwan (#110VV004).

References

1. Abbott, N.J., Patabendige, A.A., Dolman, D.E., Yusof, S.R., Begley, D.J.: Structure and function of the blood–brain barrier. Neurobiol. Dis. **37**(1), 13–25 (2010). https://doi.org/10.1016/j.nbd.2009.07.030
2. Alberding, J.P., Secomb, T.W.: Simulation of angiogenesis in three dimensions: application to cerebral cortex. PLoS Comput. Biol. **17**(6), e1009164 (2021)
3. Álvarez-Sabín, J., Maisterra, O., Santamarina, E., Kase, C.S.: Factors influencing haemorrhagic transformation in ischaemic stroke. Lancet Neurol. **12**(7), 689–705 (2013)
4. Angleys, H., Østergaard, L., Jespersen, S.N.: The effects of capillary transit time heterogeneity (CTH) on brain oxygenation. J. Cereb. Blood Flow Metab. **35**(5), 806–817 (2015)
5. Attwell, D., Mishra, A., Hall, C.N., O'Farrell, F.M., Dalkara, T.: What is a pericyte? J. Cereb. Blood Flow Metab. **36**(2), 451–455 (2016)
6. Banaji, M., Tachtsidis, I., Delpy, D., Baigent, S.: A physiological model of cerebral blood flow control. Math. Biosci. **194**(2), 125–173 (2005)
7. Beard, D.A.: Computational framework for generating transport models from databases of microvascular anatomy. Ann. Biomed. Eng. **29**(10), 837–843 (2001). https://doi.org/10.1114/1.1408920
8. Blinder, P., Tsai, P.S., Kaufhold, J.P., Knutsen, P.M., Suhl, H., Kleinfeld, D.: The cortical angiome: an interconnected vascular network with noncolumnar patterns of blood flow. Nat Neurosci. **16**(7), 889–897 (2013). https://doi.org/10.1038/nn.3426
9. Boas, D.A., Jones, S.R., Devor, A., Huppert, T.J., Dale, A.M.: A vascular anatomical network model of the spatio-temporal response to brain activation. Neuroimage **40**(3), 1116–1129 (2008)
10. Bordoni, L., Li, B., Kura, S., Boas, D.A., Sakadžić, S., Østergaard, L., Frische, S., Gutiérrez-Jiménez, E.: Quantification of capillary perfusion in an animal model of acute intracranial hypertension. J. Neurotrauma **38**(4), 446–454 (2021)
11. Brown, W.R., Thore, C.R.: Cerebral microvascular pathology in ageing and neurodegeneration. Neuropathol. Appl. Neurobiol. **37**(1), 56–74 (2011)
12. Cassot, F., Lauwers, F., Fouard, C., Prohaska, S., Lauwers-Cances, V.: A novel three-dimensional computer-assisted method for a quantitative study of microvascular networks of the human cerebral cortex. Microcirculation **13**(1), 1–18 (2006)
13. Chen, X., Józsa, T.I., Payne, S.J.: Computational modelling of cerebral oedema and osmotherapy following ischaemic stroke. Comput. Biol. Med. **151**, 106226 (2022)
14. Daher, A., Payne, S.: A network-based model of dynamic cerebral autoregulation. Microvasc. Res. **147**, 104503 (2023)
15. Deshpande, A., Jamilpour, N., Jiang, B., Michel, P., Eskandari, A., Kidwell, C., Wintermark, M., Laksari, K.: Automatic segmentation, feature extraction and comparison of healthy and stroke cerebral vasculature. NeuroImage: Clin. **30**, 102573 (2021)
16. Duvernoy, H.M., Delon, S., Vannson, J.: Cortical blood vessels of the human brain. Brain Res. Bull. **7**(5), 519–579 (1981)
17. El-Bouri, W.K., Payne, S.J.: Multi-scale homogenization of blood flow in 3-dimensional human cerebral microvascular networks. J. Theor. Biol. **380**, 40–47 (2015)

18. El-Bouri, W.K., Payne, S.J.: A statistical model of the penetrating arterioles and venules in the human cerebral cortex. Microcirculation **23**(7), 580–590 (2016)
19. El-Bouri, W.K., Payne, S.J.: Investigating the effects of a penetrating vessel occlusion with a multi-scale microvasculature model of the human cerebral cortex. NeuroImage **172**, 94–106 (2018)
20. El-Bouri, W.K., MacGowan, A., Józsa, T.I., Gounis, M.J., Payne, S.J.: Modelling the impact of clot fragmentation on the microcirculation after thrombectomy. PLoS Comput. Biol. **17**(3), e1008515 (2021)
21. Ertürk, A., Bradke, F.: High-resolution imaging of entire organs by 3-dimensional imaging of solvent cleared organs (3DISCO). Exp. Neurol. **242**, 57–64 (2013)
22. Fang, Q., Sakadzic, S., Ruvinskaya, L., Devor, A., Dale, A.M., Boas, D.A.: Oxygen advection and diffusion in a three-dimensional vascular anatomical network. Opt. Express **16**, 17530–17541 (2008)
23. Gagnon, L., Sakadžić, S., Lesage, F., Musacchia, J.J., Lefebvre, J., Fang, Q., Yücel, M.A., Evans, K.C., Mandeville, E.T., Cohen-Adad, J., et al.: Quantifying the microvascular origin of BOLD-fMRI from first principles with two-photon microscopy and an oxygen-sensitive nanoprobe. J. Neurosci. **35**(8), 3663–3675 (2015)
24. Gagnon, L., Smith, A.F., Boas, D.A., Devor, A., Secomb, T.W., Sakadžić, S.: Modeling of cerebral oxygen transport based on in vivo microscopic imaging of microvascular network structure, blood flow, and oxygenation. Front. Comput. Neurosci. **10**, 82 (2016)
25. Gkontra, P., El-Bouri, W.K., Norton, K.A., Santos, A., Popel, A.S., Payne, S.J., Arroyo, A.G.: Dynamic changes in microvascular flow conductivity and perfusion after myocardial infarction shown by image-based modeling. J. Am. Heart Assoc. **8**(7), e011058 (2019)
26. Goirand, F., Le Borgne, T., Lorthois, S.: Network-driven anomalous transport is a fundamental component of brain microvascular dysfunction. Nat. Commun. **12**(1), 7295 (2021)
27. Gould, I.G., Linninger, A.A.: Hematocrit distribution and tissue oxygenation in large microcirculatory networks. Microcirculation **22**(1), 1–18 (2015)
28. Gould, I.G., Tsai, P., Kleinfeld, D., Linninger, A.: The capillary bed offers the largest hemodynamic resistance to the cortical blood supply. J. Cereb. Blood Flow Metab. **37**(1), 52–68 (2017)
29. Graff, B.J., Payne, S.J., El-Bouri, W.K.: The ageing brain: investigating the role of age in changes to the human cerebral microvasculature with an in silico model. Front. Aging Neurosci. **13**, 632521 (2021)
30. Guibert, R., Fonta, C., Plouraboué, F.: Cerebral blood flow modeling in primate cortex. J. Cereb. Blood Flow Metab. **30**(11), 1860–1873 (2010)
31. Hall, C.N., Reynell, C., Gesslein, B., Hamilton, N.B., Mishra, A., Sutherland, B.A., O'Farrell, F.M., Buchan, A.M., Lauritzen, M., Attwell, D.: Capillary pericytes regulate cerebral blood flow in health and disease. Nature **508**(7494), 55–60 (2014)
32. Hartmann, D.A., Coelho-Santos, V., Shih, A.Y.: Pericyte control of blood flow across microvascular zones in the central nervous system. Annu. Rev. Physiol. **84**, 331–354 (2022)
33. Hartung, G., Vesel, C., Morley, R., Alaraj, A., Sled, J., Kleinfeld, D., Linninger, A.: Simulations of blood as a suspension predicts a depth dependent hematocrit in the circulation throughout the cerebral cortex. PLoS Comput. Biol. **14**(11), e1006549 (2018)
34. Hartung, G., Badr, S., Mihelic, S., Dunn, A., Cheng, X., Kura, S., Boas, D.A., Kleinfeld, D., Alaraj, A., Linninger, A.A.: Mathematical synthesis of the cortical circulation for the whole mouse brain–part ii: microcirculatory closure. Microcirculation **28**(5), e12687 (2021)
35. Hartung, G., Badr, S., Moeini, M., Lesage, F., Kleinfeld, D., Alaraj, A., Linninger, A.: Voxelized simulation of cerebral oxygen perfusion elucidates hypoxia in aged mouse cortex. PLoS Comput. Biol. **17**(1), e1008584 (2021)
36. Hashemi, Z., Rahnama, M.: Numerical simulation of transient dynamic behavior of healthy and hardened red blood cells in microcapillary flow. Int. J. Numer. Method Biomed. Eng. **32**(11), e02763 (2016)
37. Holmes, M.H.: Introduction to Perturbation Methods, vol. 20. Springer Science & Business Media, Berlin (2012)

38. Horton, N.G., Wang, K., Kobat, D., Clark, C.G., Wise, F.W., Schaffer, C.B., Xu, C.: In vivo three-photon microscopy of subcortical structures within an intact mouse brain. Nat. Photonics **7**(3), 205–209 (2013)
39. Jespersen, S.N., Østergaard, L.: The roles of cerebral blood flow, capillary transit time heterogeneity, and oxygen tension in brain oxygenation and metabolism. J. Cereb. Blood Flow Metab. **32**(2), 264–277 (2012)
40. Józsa, T.I., Padmos, R.M., El-Bouri, W.K., Hoekstra, A.G., Payne, S.J.: On the sensitivity analysis of porous finite element models for cerebral perfusion estimation. Ann. Biomed. Eng. **49**(12), 3647–3665 (2021)
41. Józsa, T.I., Padmos, R.M., Samuels, N., El-Bouri, W., Hoekstra, A.G., Payne, S.J.: A porous circulation model of the human brain for in silico clinical trials in ischaemic stroke. Interface Focus **11**(1), 20190127 (2021)
42. Krüger, T., Varnik, F., Raabe, D.: Efficient and accurate simulations of deformable particles immersed in a fluid using a combined immersed boundary lattice Boltzmann finite element method. Comput. Math. Appl. **61**(12), 3485–3505 (2011)
43. Linninger, A., Gould, I., Marinnan, T., Hsu, C.Y., Chojecki, M., Alaraj, A.: Cerebral microcirculation and oxygen tension in the human secondary cortex. Ann. Biomed. Eng. **41**, 2264–2284 (2013)
44. Linninger, A., Hartung, G., Badr, S., Morley, R.: Mathematical synthesis of the cortical circulation for the whole mouse brain-part i. Theory and image integration. Comput. Biol. Med. **110**, 265–275 (2019)
45. Lorthois, S., Cassot, F., Lauwers, F.: Simulation study of brain blood flow regulation by intra-cortical arterioles in an anatomically accurate large human vascular network: Part i: methodology and baseline flow. NeuroImage **54**, 1031–1042 (2011). https://doi.org/10.1016/j.neuroimage.2010.09.032
46. Lorthois, S., Cassot, F., Lauwers, F.: Simulation study of brain blood flow regulation by intra-cortical arterioles in an anatomically accurate large human vascular network. part ii: flow variations induced by global or localized modifications of arteriolar diameters. NeuroImage **54**(4), 2840–2853 (2011)
47. Marchand, P.J., Lu, X., Zhang, C., Lesage, F.: Validation of red blood cell flux and velocity estimations based on optical coherence tomography intensity fluctuations. Sci. Rep. **10**(1), 1–10 (2020)
48. Milanovic, S., Shaw, K., Hall, C., Payne, S.: Investigating the role of pericytes in cerebral autoregulation: a modeling study. Physiol. Meas. **42**(5), 054003 (2021)
49. Moeini, M., Lu, X., Avti, P.K., Damseh, R., Bélanger, S., Picard, F., Boas, D., Kakkar, A., Lesage, F.: Compromised microvascular oxygen delivery increases brain tissue vulnerability with age. Sci. Rep. **8**(1), 8219 (2018)
50. Mokhtarudin, M.J.M., Payne, S.J.: The study of the function of aqp4 in cerebral ischaemia–reperfusion injury using poroelastic theory. Int. J. Numer. Method Biomed. Eng. **33**(1), e02784 (2017)
51. Mokhtarudin, M.J.M., Naim, W.N.W.A., Shabudin, A., Payne, S.J.: Multiscale modelling of brain tissue oxygen and glucose dynamics in tortuous capillary during ischaemia-reperfusion. Appl. Math. Model. **109**, 358–373 (2022)
52. Padmos, R.M., Józsa, T.I., El-Bouri, W.K., Konduri, P.R., Payne, S.J., Hoekstra, A.G.: Coupling one-dimensional arterial blood flow to three-dimensional tissue perfusion models for in silico trials of acute ischaemic stroke. Interface Focus **11**(1), 20190125 (2021)
53. Park, C.S., Payne, S.J.: A generalized mathematical framework for estimating the residue function for arbitrary vascular networks. Interface Focus **3**(2), 20120078 (2013)
54. Payne, S.J.: Cerebral Blood Flow and Metabolism: A Quantitative Approach. World Scientific, Singapore (2018)
55. Payne, S., Józsa, T.I., El-Bouri, W.K.: Review of in silico models of cerebral blood flow in health and pathology. Prog. Biomed. Eng. **5**(2), 022003 (2023)
56. Penta, R., Gerisch, A.: Investigation of the potential of asymptotic homogenization for elastic composites via a three-dimensional computational study. Comput. Vis. Sci. **17**, 185–201 (2015)

57. Peyrounette, M., Davit, Y., Quintard, M., Lorthois, S.: Multiscale modelling of blood flow in cerebral microcirculation: details at capillary scale control accuracy at the level of the cortex. PLoS One **13**(1), e0189474 (2018)
58. Pias, S.C.: How does oxygen diffuse from capillaries to tissue mitochondria? Barriers and pathways. J. Physiol. **599**(6), 1769–1782 (2021)
59. Price, G.F., Chernyavsky, I.L., Jensen, O.E.: Advection-dominated transport past isolated disordered sinks: stepping beyond homogenization. Proc. R. Soc. A **478**(2262), 20220032 (2022)
60. Pries, A.R., Secomb, T.W.: Microvascular blood viscosity in vivo and the endothelial surface layer. Am. J. Physiol. Heart Circ. Physiol. **289**(6), H2657–H2664 (2005)
61. Pries, A., Secomb, T., Gaehtgens, P., Gross, J.: Blood flow in microvascular networks. Experiments and simulation. Circ. Res. **67**(4), 826–834 (1990)
62. Reichold, J., Stampanoni, M., Keller, A.L., Buck, A., Jenny, P., Weber, B.: Vascular graph model to simulate the cerebral blood flow in realistic vascular networks. J. Cereb. Blood Flow Metab. **29**(8), 1429–1443 (2009)
63. Riddle, D.R., Sonntag, W.E., Lichtenwalner, R.J.: Microvascular plasticity in aging. Ageing Res. Rev. **2**(2), 149–168 (2003)
64. Safaeian, N., David, T.: A computational model of oxygen transport in the cerebrocapillary levels for normal and pathologic brain function. J. Cereb. Blood Flow Metab. **33**(10), 1633–1641 (2013)
65. Schmid, F., Tsai, P.S., Kleinfeld, D., Jenny, P., Weber, B.: Depth-dependent flow and pressure characteristics in cortical microvascular networks. PLoS Comput. Biol. **13**(2), e1005392 (2017)
66. Schmid, F., Barrett, M.J., Jenny, P., Weber, B.: Vascular density and distribution in neocortex. NeuroImage **197**, 792–805 (2019)
67. Schneider, M., Reichold, J., Weber, B., Székely, G., Hirsch, S.: Tissue metabolism driven arterial tree generation. Med. Image Anal. **16**(7), 1397–1414 (2012)
68. Secomb, T.W., Hsu, R., Beamer, N.B., Coull, B.M.: Theoretical simulation of oxygen transport to brain by networks of microvessels: effects of oxygen supply and demand on tissue hypoxia. Microcirculation **7**(4), 237–247 (2000)
69. Sharma, D., Smith, M.: The intensive care management of acute ischaemic stroke. Curr. Opin. Crit. Care **28**(2), 157–165 (2022)
70. Shaw, K., Bell, L., Boyd, K., Grijseels, D., Clarke, D., Bonnar, O., Crombag, H., Hall, C.: Neurovascular coupling and oxygenation are decreased in hippocampus compared to neocortex because of microvascular differences. Nat. Commun. **12**(1), 3190 (2021)
71. Smith, A.F., Doyeux, V., Berg, M., Peyrounette, M., Haft-Javaherian, M., Larue, A.E., Slater, J.H., Lauwers, F., Blinder, P., Tsai, P., et al.: Brain capillary networks across species: a few simple organizational requirements are sufficient to reproduce both structure and function. Front. Physiol. **10**, 233 (2019)
72. Su, S.W., Catherall, M., Payne, S.: The influence of network structure on the transport of blood in the human cerebral microvasculature. Microcirculation **19**(2), 175–187 (2012)
73. Tahir, A.M., Chowdhury, M.E., Khandakar, A., Rahman, T., Qiblawey, Y., Khurshid, U., Kiranyaz, S., Ibtehaz, N., Rahman, M.S., Al-Maadeed, S., et al.: Covid-19 infection localization and severity grading from chest x-ray images. Comput. Biol. Med. **139**, 105002 (2021)
74. Tong, Z., Catherall, M., Payne, S.J.: A multiscale model of cerebral autoregulation. Med. Eng. Phys. **95**, 51–63 (2021)
75. Tsai, P.S., Friedman, B., Ifarraguerri, A.I., Thompson, B.D., Lev-Ram, V., Schaffer, C.B., Xiong, Q., Tsien, R.Y., Squier, J.A., Kleinfeld, D.: All-optical histology using ultrashort laser pulses. Neuron **39**(1), 27–41 (2003)
76. Tsai, P.S., Kaufhold, J.P., Blinder, P., Friedman, B., Drew, P.J., Karten, H.J., Lyden, P.D., Kleinfeld, D.: Correlations of neuronal and microvascular densities in murine cortex revealed by direct counting and colocalization of nuclei and vessels. J. Neurosci. **29**(46), 14553–14570 (2009)

77. Ursino, M., Lodi, C.A.: A simple mathematical model of the interaction between intracranial pressure and cerebral hemodynamics. J. Appl. Physiol. **82**(4), 1256–1269 (1997)
78. van der Wijk, A.E., Georgakopoulou, T., Majolée, J., van Bezu, J.S., van der Stoel, M.M., van Het Hof, B.J., de Vries, H.E., Huveneers, S., Hordijk, P.L., Bakker, E.N., et al.: Microembolus clearance through angiophagy is an auxiliary mechanism preserving tissue perfusion in the rat brain. Acta Neuropathol. Commun. **8**(1), 195 (2020)
79. Van Rooij, B., Závodszky, G., Azizi Tarksalooyeh, V., Hoekstra, A.: Identifying the start of a platelet aggregate by the shear rate and the cell-depleted layer. J. R. Soc. Interface **16**(159), 20190148 (2019)
80. Ventimiglia, T., Linninger, A.A.: Mesh-free high-resolution simulation of cerebrocortical oxygen supply with fast Fourier preconditioning. Int. J. Numer. Method Biomed. Eng. **39**, e3735 (2023)
81. Walsh, C., Tafforeau, P., Wagner, W., Jafree, D., Bellier, A., Werlein, C., Kühnel, M., Boller, E., Walker-Samuel, S., Robertus, J., et al.: Imaging intact human organs with local resolution of cellular structures using hierarchical phase-contrast tomography. Nat. Methods **18**(12), 1532–1541 (2021)
82. Wang, J., Payne, S.J.: Mathematical modelling of haemorrhagic transformation after ischaemic stroke. J. Theor. Biol. **531**, 110920 (2021)
83. Wang, J., Van Kranendonk, K.R., El-Bouri, W.K., Majoie, C.B., Payne, S.J.: Mathematical modelling of haemorrhagic transformation within a multiscale microvasculature network. Physiol. Meas. **43**(5), 055006 (2022)
84. Weber, B., Keller, A.L., Reichold, J., Logothetis, N.K.: The microvascular system of the striate and extrastriate visual cortex of the macaque. Cereb. Cortex **18**(10), 2318–2330 (2008)
85. Xue, S., Gong, H., Jiang, T., Luo, W., Meng, Y., Liu, Q., Chen, S., Li, A.: Indian-ink perfusion based method for reconstructing continuous vascular networks in whole mouse brain. PLoS One **9**(1), e88067 (2014)
86. Xue, Y., El-Bouri, W.K., Józsa, T.I., Payne, S.J.: Modelling the effects of cerebral microthrombi on tissue oxygenation and cell death. J. Biomech. **127**, 110705 (2021)
87. Zagzoule, M., Marc-Vergnes, J.P.: A global mathematical model of the cerebral circulation in man. J. Biomech. **19**(12), 1015–1022 (1986)
88. Závodszky, G., Van Rooij, B., Azizi, V., Hoekstra, A.: Cellular level in-silico modeling of blood rheology with an improved material model for red blood cells. Front. Physiol. **8**, 563 (2017)

A Computational Model of the Tumor Microenvironment Applied to Fractionated Radiotherapy

Luca Possenti, Andrea Gallo, Piermario Vitullo, Alessandro Cicchetti, Tiziana Rancati, Maria Laura Costantino, and Paolo Zunino

Abstract Radiotherapy consists in delivering a precise radiation dose to a specific tumor target in order to eradicate tumor cells and to achieve local tumor control. The definition of the most suitable radiotherapy treatment schedule is not trivial due to the large tumor heterogeneity reported in clinical practice. The ultimate goal is to prescribe a specific treatment pattern for each patient, considering all the different radiobiological properties of the tumor/normal tissues to achieve the best final result possible. The model presented in this work goes in this direction, analyzing oxygen dependency and the role of the vascular network in the tumor microenvironment, since the efficacy of radiation therapy also depends on local oxygen availability. The main purpose of this work is to develop a mathematical model that describes the interaction between microvascular oxygen transfer and the efficacy of fractionated radiotherapy.

Keywords Oxygen transfer · Microvascular environment · Radiotherapy · Computational modelling

---

L. Possenti · A. Cicchetti · T. Rancati
Fondazione IRCCS Istituto Nazionale dei Tumori, Milano, Italy
e-mail: luca.possenti@istitutotumori.mi.it; alessandro.cicchetti@istitutotumori.mi.it; tiziana.rancati@istitutotumori.mi.it

A. Gallo · M. L. Costantino
Department of Chemistry, Materials and Chemical Engineering "Giulio Natta", Politecnico di Milano, Milano, Italy
e-mail: andrea5.gallo@mail.polimi.it; marialaura.costantino@polimi.it

P. Vitullo · P. Zunino (✉)
MOX, Department of Mathematics, Politecnico di Milano, Milano, Italy
e-mail: piermario.vitullo@polimi.it; paolo.zunino@polimi.it

A. Linninger et al. (eds.), *Quantitative Approaches to Microcirculation*,
SEMA SIMAI Springer Series 36, https://doi.org/10.1007/978-3-031-58519-7_2

1 Introduction

Radiotherapy is recommended as part of treatment for more than 50% cancer patients [12]. It delivers a precise radiation dose to a specific tumor target to eradicate clonogenic (tumor) cells and achieve local tumor control. Identifying the most appropriate radiotherapy treatment schedule is not trivial due to the large tumor heterogeneity reported in clinical practice. Different tumor aggressiveness, tissue and tumor sensibility to radiation (i.e., radiosensitivity), and also possible microscopic/macroscopical tumor spread raise the level of complexity in radiotherapy treatment planning [5]. The ideal scenario would be the prescription of a specific treatment pattern for each patient, considering all radiobiological properties of the tumor/normal tissues and achieving the best final outcome possible. The model presented in this work moves exactly in this direction through the analysis of two specific factors that affect radiotherapy: oxygen dependency and the role of the vascular network.

Modern treatment planning guidelines in radiotherapy include dose fractionation [33], which consists of dividing the total dose into a set of daily dose fractions, with the purpose of minimizing side effects. The choice of the specific schedule (i.e., how many fractions, the amount of dose in each fraction, and the time gap between consecutive fractions) is driven by the radiobiological properties of the tumor and the normal tissues surrounding the target [8]. Normal tissue generally benefits from a high number of small-dose fractions, and tumors are somewhat less sensitive to fractionation.

Individual treatment planning and dose fractionation are determined by many factors, including those provided by radiobiological models, which describe the effect of radiation on cells and tissue [43]. When cells are irradiated, the integrity of the DNA structure is affected, and in some cases, huge modifications may appear in the genome. Radiation exposure can directly cause the rupture of chemical bonds that build up DNA or produce indirect damage by ionization of some molecules (e.g. water, with the formation of free radicals) that will enter in contact with DNA and cause chain breaks. At this level, the oxygen present in the vascular microenvironment plays a crucial role, stabilizing the radicals already formed, and thus securing the damage [9]. In case of low oxygen concentration, this process is impaired, leading to radioresistant behavior [14]. To evaluate the efficacy of radiotherapy, the concept of cell surviving fraction (SF) is commonly adopted, as it quantifies the proportion of cells that survive exposure to doses. Its value is a number in the range of 0 to 1, often expressed as a percentage. Different radiobiological models have been proposed to accurately predict SF before starting the radiotherapy treatment. Among them, the Linear-Quadratic (LQ) model is the gold standard, thanks to its simplicity and prediction capabilities [25]. It assumes a proportionality between cell SF and the dose delivered, expressing the absorbed energy per unit mass of tissue (measured in Gray [Gy = J/kg] in the SI system).

In this work, we consider a computational model to analyze the spatial distribution of the partial pressure of oxygen, or concentration, in the vascular microenvironment, deducing from different scenarios the importance of biological, mechanical and fluid parameters in the consequent hypoxia condition. Microvascular oxygen delivery is typically analyzed computationally, reducing the dimensionality of the vasculature by utilizing different methods. Models that account for the local morphology of the microvasculature have been developed recently, describing the microvasculature as a collection of localized sources, see for example [39] and more recently [13, 42]. This seminal idea has evolved into advanced multiphysics approaches [21, 31, 32], embracing vascular and interstitial flow, red blood cell transport, oxygen transport, angiogenesis, and enabling mesoscale analysis of the vascular microenvironment.

However, few of these models have been applied to study RT and tumor microenvironment. Scott and colleagues described the interaction between vascular density and a morphological index [38]. In a subsequent study, the same authors showed the importance of microvascular oxygen transport [11].

The main purpose of this work is to bridge the gap between advanced models of the vascular microenvironment and the application of radiobiological models to formulate fractionated radiotherapy plans. We achieve this goal using a mathematical model that describes the interaction between microvascular oxygen transfer and the efficacy of fractionated radiotherapy.

2 An Oxygen Transport Model Coupled to Radiotherapy

In this section, partial differential equations (PDE) are used to model the physics of transport phenomena in the vascular microenvironment. Then the model is coupled with a radiobiological model to include the effect of radiotherapy.

2.1 A Mixed-Dimensional Oxygen Transport Model

The mathematical framework of the vascular microenvironment is inherited from previous work by Possenti et al., where oxygen transport in a piece of tissue was modeled [30, 31]. In this chapter, the basic equations of the model are described in order to provide a context for the coupling with radiotherapy. The quantity of interest in this microcirculation problem is the oxygen concentration, transported both as free oxygen concentration (C) and bounded to hemoglobin ($CHbO_2$). Hemoglobin is a macromolecule incorporated in red blood cells (RBCs), which can chemically bond with oxygen molecules. In this way, oxygen is transported through the blood vessel thanks to the red blood cells. Furthermore, oxygen could diffuse out of the RBC and then out of the vessel, crossing the vascular wall. For this reason, two regions are identified: Ω_t, corresponding to the tissue, and Ω_v, related to the vascular network (plasma and RBCs). The geometry of the vessels is reduced to a 1D object, identified by Λ [23, 31]. In the transport problem, oxygen concentration is treated as a function of both space and time: $C = C(x, t)$. Depending on the physical domain in which C is defined, a different notation is used for tissue or vessel. If the oxygen concentration in the tissue is considered, we use C_t and measure it using measured using $\text{ml}_{O_2}/\text{ml}_{tissue}$. On the contrary, the oxygen concentration in the vessel is indicated as C_v, where a different set of units is adopted, namely $\text{ml}_{O_2}/\text{ml}_{blood}$. Furthermore, two main assumptions are included, back from the previous work [31]:

- Steady state condition: the model does not consider time derivatives. It is reasonable to assume transport phenomena in microcirculation to occur in a slow-paced way, neglecting any transient variation to the physical quantities involved;
- Absence of body forces: no body forces (i.e. gravity, inertia) are included in this work.

We start with blood flow in the microvascular network. A mass balance is established for the transport of oxygen across the capillary membrane. In this scenario, it is important to note that oxygen is present as a free concentration as well as bound to hemoglobin $C_{tot} = C_v + C_{HbO_2}$, which must be considered. Moreover, two assumptions are required:

- Oxyhemoglobin cannot diffuse through the membrane: once the oxygen is bound to hemoglobin (inside RBC) the oxyhemoglobin molecule is successfully created. This molecule cannot diffuse outside the RBCs and therefore through the capillary membrane because of its large size;
- Temporal dynamics of oxygen diffusion through RBC is neglected: when oxygen leaves or enters the RBC, a diffusion process takes place. However, this kind of phenomenon is considered instantaneous in this framework, so it will not be relevant to the modeling.

Then, the general 3D transport equation of oxygen in the vessel could be written as:

$$\pi R^2 \, D_v \frac{\partial^2 C_v}{\partial s^2} - \pi R^2 \frac{\partial (u_v \, C_{tot})}{\partial s} = -J_{O_2}. \tag{1}$$

The main mechanisms that regulate oxygen transport in the tissue surrounding the vascular network are the following:

- Diffusion: this term describes oxygen diffusion in Ω_t. It is mainly governed by the diffusion coefficient D_t, expressed in m^2/s;
- Advection: it represents the motion of oxygen by convection, regulated by velocity u_t, expressed in m/s;
- Reaction: a sink term, illustrated by M, the consumption coefficient. It accounts for the amount of oxygen consumed by the tissue and is measured in ml O_2/cm^3.

Based on the aforementioned assumptions, the transport equation in the tissue reads:

$$\nabla \cdot (D_t \nabla (C_t)) - \nabla \cdot (\mathbf{u_t} \, C_t) + M \, C_t = J_{O_2} \quad \text{on } \Omega_t. \tag{2}$$

Finally, the oxygen transfer model is completed by describing the mechanisms that govern the tissue-vessel interaction. This model describes the total mass flow of oxygen J_{O_2}, which accounts for the sum of the diffusive mass flow (J_{diff}) and convective (J_{adv}) mass flow. Exploiting the theory of passive transport through semipermeable membranes, this term is modeled thanks to the Kedem-Katchalsky equation [17] as:

$$\begin{cases} J_{diff} = 2\pi R \ P_L(C_v - C_t) \\ J_{adv} = 2\pi R \left(\dfrac{C_v + C_t}{2}\right) L_p \ (\Delta P - \sigma \Delta\pi) \\ J_{O_2} = J_{diff} + J_{adv}, \end{cases} \tag{3}$$

where L_p is the hydraulic conductivity of the microvascular wall and π is the osmotic pressure and P_L is the vascular wall permeability to oxygen. We point out that, for small molecules, diffusion across the endothelial layer dominates greatly over transport. For example, using the parameters reported in Table 1 for oxygen, we see that the coefficient $2\pi R \ P_L$ of J_{diff} is four orders of magnitude higher than $2\pi R \ L_p$ appearing in J_{adv}. The second term on the right-hand side of (3) is important if the semi-permeable model is used for the transfer of large molecules, such as proteins, or to model very leaky vessels that can appear in the case of tumors.

2.2 *Constitutive Laws of the Oxygen Transport Model*

The model previously described is complemented by the following (non-linear) constitutive laws, namely, the Michaelis-Menten equation and the oxyhemoglobin concentration. The former describes the non-linear behavior of the oxygen consumption rate:

$$M(pO_2) = V_{max} \frac{pO_2}{pO_2 + P_{m_{50}}}, \tag{4}$$

where the consumption rate M is described as a function of oxygen partial pressure pO_2. The other two parameters are the maximum consumption rate in the tissue V_{max} (mlO_2/cm^3/s) and the partial pressure of oxygen at half the consumption rate $P_{m_{50}}$ (mmHg). Henry's law is applied ($C = \kappa P$), where a linear relationship between gas concentration and partial pressure is assumed. The linear parameter introduced (κ) represents the gas solubility in the environment and is a function of temperature ($\kappa = \kappa(T)$). So, Eq. (4) is modified as:

$$M(C_{O_2}) = V_{max} \frac{C_{O_2}}{C_{O_2} + \kappa_t \ P_{m_{50}}}. \tag{5}$$

For V_{max} and $P_{m_{50}}$, physiological values from the literature will be adopted ($V_{max} = 6.17 \cdot 10^{-5}$ $\text{ml}_{O_2}/\text{cm}^3/\text{s}$, $P_{m_{50}} = 1.0$ mmHg) [32]. The oxygen solubility in the interstitium ($\kappa_t = 3.89 \cdot 10^{-5}$ $\text{ml}_{O_2}/\text{cm}^3/\text{mmHg}$) is introduced and when coupled with $P_{m_{50}}$ gives rise to the Michaelis-Menten constant K_M $(= \kappa_t\ P_{m50})$.

The nonlinearity of the oxyhemoglobin concentration is described by the Hill equation to account for the saturation of RBC. In fact, the overall oxyhemoglobin concentration could be defined as $C_{HbO_2} = C_{Hb}\ S(pO_2)$ that involves the concentration of hemoglobin in the blood (C_{Hb}) and the saturation of hemoglobin within the RBC, named $S(pO_2)$. Furthermore, hemoglobin concentration depends on the binding capacity of heme groups N, the amount of hematocrit (H_t) in the blood (number of RBCs), and the Mean Corpuscular Hemoglobin Concentration ($MCHC$), representing the concentration of Hb in each RBC. As a result, we obtain $C_{Hb} = k_1\ H_t$, where $k_1 = N\ MCHC$. Oxygen saturation can be modeled using the Hill equation as follows:

$$S(pO_2) = \frac{pO_2^{\gamma}}{pO_2^{\gamma} + P_{s50}^{\gamma}}. \tag{6}$$

The formulation is similar to the Michaelis-Menten formula (Eq. (4)), except for the Hill constant γ and $P_{s_{50}}$, the partial pressure of oxygen when hemoglobin reaches half saturation. Thus, applying Henry's law once again ($C = \kappa\ P$):

$$C_{HbO_2} = k_1 \cdot H_t \cdot \frac{C_{O_2}^{\gamma}}{C_{O_2}^{\gamma} + k_2^{\gamma}} \tag{7}$$

where $k_2 = (\kappa_{pl}\ P_{s_{50}})^{\gamma}$ is introduced to simplify the notation. Table 1 summarizes the values adopted in this work, obtained by fitting the Hill curve under physiological conditions [36].

Including the non-linear terms (Eqs. (7) and (5)) in the transport problem (Eq. (2)), the overall system reads as follows:

$$\begin{cases} \nabla \cdot (D_t \nabla(C_t)) - \nabla \cdot (\mathbf{u_t}\ C_t) + V_{max}\ \dfrac{C_t}{C_t + K_M} = \\ 2\pi R\ [P_L(C_v - C_t) + \left(\dfrac{C_v + C_t}{2}\right) L_p\ (\Delta P - \sigma\, \Delta\pi)] \quad \text{on } \Omega_t \\ \\ \pi R^2\ D_v \dfrac{\partial^2 C_v}{\partial s^2} - \pi R^2 \dfrac{\partial(\mathbf{u_v}\ C_v)}{\partial s} - \pi R^2 \dfrac{\partial}{\partial s}\left(\mathbf{u_v}\ k_1\ H_t\ \dfrac{C_v^{\gamma}}{C_v^{\gamma} + k_2}\right) = \\ -2\pi R\ [P_L(C_v - C_t) + \left(\dfrac{C_v + C_t}{2}\right) L_p\ (1 - \sigma_{oxy})\ (\Delta P - \sigma\, \Delta\pi)] \quad \text{on } \Lambda. \end{cases} \tag{8}$$

2.3 The Linear-Quadratic Radiobiological Model

Radiotherapy consists in delivering a precise radiation dose to a specific tumor target in order to control the proliferation of clonogenic cells and possibly eradicate them. Evaluation of the cell death rate becomes crucial in this regard. Through clonogenic assays, for example, it is possible to quantify in-vitro the number of cells before and after radiation exposure, ensuring the clinical outcome prescribed by clinicians. To this purpose, the cell surviving fraction (SF) is evaluated, consisting of the proportion of cells that survive exposure to dose (i.e. cells that can still form colonies), usually computed as a percentage.

The eradication of cells described above is directly proportional to the total dose administered. Clinicians generally prescribe a sufficient dose to completely eradicate a tumor (SF close to 0), but the dose is typically not administered all at once. Indeed, the concept of fractionation is adopted, consisting of subdividing the dose into a set of dose fractions generally administered in time according to the radiobiological properties of the tumor and the surrounding tissues involved. The use of fractionation is justified by the 4R principle of radiobiology, demonstrating that splitting the dose into fractions brings benefits in terms of *repair, repopulation, redistribution, and reoxygenation* [26]. Tumors (and normal tissues) are dynamic elements constantly subjected to a very intense biological activity. Fractionation gives the possibility to normal tissue (slow proliferation, late responding cell) to repair part of the damage from one fraction to the next, while tumors (high proliferation, acute responding cells) are not affected by this process. Furthermore, improvement in terms of reoxygenation in tumors after every single fraction is related to vascular reorganization after vascular damage, enhancing oxygen redistribution in the tumor, and thus increasing the efficacy of radiotherapy in the next fraction. Due to these biological properties, an adequate and well-designed fractionation can widen the "therapeutic window", that is, increase the probability of eradicating the tumor without affecting the surrounding normal tissues [37].

The use of mathematical models in radiobiology has been well established since the past century, when the first type of model was developed to help investigators analyze and predict the effect of radiation exposure [4]. The Linear-Quadratic (LQ) model is the gold standard due to its simplicity and prediction capabilities. In this model, a lethal event in a single cell (single energy deposition causing a double-strand break in DNA) is taken into account by the linear term (α), whereas a sequence of multiple events causing a double-strand break (i.e. interaction of multiple radiation tracks) is demanded by the second-order term (β). The model was first described by Sinclair et al. in 1966 when the aim was to find a mathematical expression to fit the experimental data [41]; then Kellerer and Rossi (1971) formally derived the LQ model framework from the previous studies in single- and multihit models [18]. The formal equation of the LQ model is the following:

$$SF(D) = \exp(-\alpha D - \beta D^2). \tag{9}$$

This equation shows the surviving fraction (SF) as a function of the total dose provided (D) through some general parameters (α, β), representing the radiosensitivity of the tissue. The SF value is a number in the range of 0 to 1 (also expressed as a percentage) and is represented on a logarithmic scale. The formalism used is the standard, where a complete repair of sublethal damage between one fraction and the following is assumed. The linear parameter α [Gy^{-1}] models the lethal damage provided to cells after one single hit, so it is linearly related to the dose. The quadratic parameter β [Gy^{-2}] instead involves the accumulation of damage after multiple hits, proportional to the square of the dose administered. The value of these parameters depends on the ability of the tissue to resist radiation exposure, according to the type of cells involved and other tissue properties.

2.4 Coupling the Oxygen Transport and the Radiobiological Models

Describing the effect of radiotherapy on vascularized tissue is a very complex task [1, 44]. Radiation can start and accelerate atherosclerosis, leading to problems such as stroke and heart disease. In the tumor microenvironment, exposure to radiation has many effects: damage DNA, affect small blood vessels in both tumor and healthy tissue, and alter the way the immune system responds. Specifically, radiation harms endothelial cells (cells that line blood vessels). They become leaky, detach from their usual place, and can die. This dysfunction and cell death can cause inflammation, scarring, and blood clotting after radiation. Although changes caused by radiotherapy depend on the dose and type of radiation, predicting exactly how the TME will change is still a great challenge. Additionally, radiation can damage small blood vessels outside the tumor. The severity of the damage depends on the importance of the organ or system involved. Responses of normal tissues to radiation, both immediate and delayed, can limit the dose used in RT and affect the effectiveness of treatment and the patient's quality of life. It is vital to understand how radiation impacts small blood vessels in the tumor, but it has been difficult to study because of the challenges of watching how all these components respond together over time and in a living body. Traditional statistical models, which are heavily based on collecting a lot of data and finding associations between variables and outcomes, are not good at describing this complex environment. In conclusion, when considering vascular damage, its implications remain unclear, especially in a manner that facilitates integration within a model. Similarly, this claim is applicable to the remodeling of the vascular network. Currently, there is a paucity of data to elucidate how this occurs without resorting to substantial assumptions.

For these reasons, we present here a simplified model that accounts for the effect of radiotherapy on the vascular microenvironment, with the aim of coupling radiotherapy with oxygen transport. This model only addresses the effect of radiotherapy on the oxygen consumption rate, due to the variation in the number of cells in the tissue after radiotherapy. Therefore, we obtain different oxygen distributions at each fractionation step. In practice, we choose to modify the Michaelis-Menten formula for oxygen consumption. Therefore, the reaction term is modified as shown:

$$M(C_t) = V_{max} \frac{C_t}{C_t + K_M} SF(C_t) \tag{10}$$

introducing a multiplication by the surviving fraction (SF). As the number of viable cells decreases during treatment, the consumption rate M also decreases, affecting the simulation of C_t. This effect is relevant only in the tissue domain (Ω_t), since, as discussed above, radiation damage to the vessels is not addressed in this work. Furthermore, we neglect the possible dynamic process of cell death (e.g., death via a later mitotic catastrophe), assuming that cell death occurs before the subsequent fraction.

To calculate SF we start from the classic linear-quadratic model (9). According to Wenzl and colleagues, α and β could be directly calculated, introducing the effect of the partial pressure of oxygen and the linear energy transfer parameter (LET), namely the amount of energy transferred per unit distance, depending on the type of radiation and its features [48, 49]. The radiosensitivity parameters are defined as,

$$\begin{cases} \alpha\,(pO_2, LET) = \dfrac{(a_1 + a_2\,LET)\cdot pO_2 + (a_3 + a_4\,LET)\cdot K}{pO_2 + K} \\ \sqrt{\beta\,(pO_2)} = \dfrac{b_1\,pO_2 + b_2\,K}{pO_2 + K} \end{cases} \tag{11}$$

where $a_1, a_2, a_3, a_4, b_1, b_2$ are parameters obtained by fitting *in vitro* data [48]. Instead, the coefficient K, is the oxygen tension at which the relative radiosensitivity reaches half of its maximum value and in this case is assumed equal to 2.5 mmHg [36]. As usual, each equation written in terms of oxygen partial pressure is rewritten as a function of the oxygen concentration thanks to the Henry's law ($pO_2\,(x) = C_t\,(x)/\kappa_t$). Thus, the surviving fraction reads as follows:

$$\begin{cases} SF\,(C_t) = exp\left(-\alpha(C_t)\,D - \beta(C_t)\,D^2\right) \\ \alpha\,(C_t, LET) = \dfrac{(a_1 + a_2\,LET)\cdot \dfrac{C_t}{\kappa_t} + (a_3 + a_4\,LET)\cdot K}{\dfrac{C_t}{\kappa_t} + K} \\ \sqrt{\beta\,(C_t)} = \dfrac{b_1\,\dfrac{C_t}{\kappa_t} + b_2\,K}{\dfrac{C_t}{\kappa_t} + K} \end{cases} \tag{12}$$

As mentioned above, fractionation consists of subdividing the total dose prescribed to the patient into smaller doses, called "fractions", in order to reduce normal tissue damage to the surroundings. To include the concept of fractionation, the fractionation index z is introduced. The fractionation index describes the distribution of dose fractions in the tumor and, therefore, the analysis of different fractionation patterns adopted by clinicians. The resulting SF will be evaluated fraction by fraction, summing the effect of those previously calculated. So, the fractionation index is defined as $z = 1, 2, 3, .., n_z$ ($z \in \mathbb{N}$), where n_z is the total number of fractions in which the total dose is subdivided. The counter on z progresses with the number of fractions effectively delivered. The fractionation index is initially set equal to 1 because at least one dose fraction is supposed to be released.

In this setting, the linear/quadratic model applied to a sequence of radiotherapy sessions, i.e. a fractionated treatment, is obtained from Eq. (12) as follows:

$$\begin{cases} SF^{(z)}\left(C_t\right) = SF^{(z-1)}\, exp\left[\left(-\alpha^{(z)} D - \beta^{(z)} D^2\right)\right] \\ \alpha^{(z)}\left(C_t, LET\right) = \dfrac{\left(a_1 + a_2 LET\right) \cdot \dfrac{C_t^{(z-1)}}{K_t} + \left(a_3 + a_4 LET\right) \cdot K}{\dfrac{C_t^{(z-1)}}{K_t} + K} \\ \sqrt{\beta^{(z)}\left(C_t\right)} = \dfrac{b_1 \dfrac{C_t^{(z-1)}}{K_t} + b_2 K}{\dfrac{C_t^{(z-1)}}{K_t} + K} \end{cases} \tag{13}$$

The transition between the previous radiation exposure, indicated as $z - 1$, and the current one, z, determines the dose accumulation process. Initial conditions are required to exploit the capability of the model fully:

- $SF^{(0)}$ must be set equal to 1 as it corresponds to the non-irradiation condition;
- $C_t^{(0)}$ corresponds to the oxygen distribution before the treatment starts. It is computed by assuming $SF^{(0)}$.

At this point, we are ready to couple the vascular microenvironment model with the one for fractionated radiotherapy. Starting from (8), we replace (10) with the original Michaelis-Menten term, $V_{max}\ \frac{C_t}{C_t + K_M}$. Then we express the survival fraction after each radiotherapy session as (13). In conclusion, the coupled model becomes:

$$
\begin{cases}
\nabla \cdot (D_t \nabla (C_t^{(z)})) - \nabla \cdot (\mathbf{u_t}\, C_t^{(z)}) + V_{max}\, \dfrac{C_t^{(z)}}{C_t^{(z)} + K_M}\, SF^{(z-1)} = \\
\quad 2\pi R\, [P_L (C_v^{(z)} - C_t^{(z)}) \\
\quad + (\dfrac{C_v^{(z)} + C_t^{(z)}}{2}) L_p\, (1 - \sigma_{oxy})\, (\Delta P - \sigma \Delta\pi)] \quad \text{on } \Omega_t \\
\pi R^2\, D_v \dfrac{\partial^2 C_v^{(z)}}{\partial s^2} - \pi R^2 \dfrac{\partial (\mathbf{u_v}\, C_v^{(z)})}{\partial s} - \pi R^2 \dfrac{\partial}{\partial s} (\mathbf{u_v}\, k_1\, H_t\, \dfrac{C_v^{(z),\gamma}}{C_v^{(z),\gamma} + k_2}) = \\
\quad -2\pi R\, [P_L (C_v^{(z)} - C_t^{(z)}) \\
\quad + (\dfrac{C_v^{(z)} + C_t^{(z)}}{2}) L_p\, (\Delta P - \sigma \Delta\pi)] \quad \text{on } \Lambda
\end{cases}
\tag{14}
$$

The model (14) represents a sequence of states of the vascular microenvironment during fractionated radiotherapy based on a quasi-static modeling approach. The simulation of a fractionated therapy requires an iterative process that describes how the tissue oxygenation and the survival fraction evolve from one fraction to the next one through the following steps:

- solve for the oxygen concentration in the tissue C_t and in the vessel C_v;
- update the value of SF, function of C_t obtained at the previous step;
- move to the next dose fraction, where SF will directly influence the oxygen distribution;
- back to the start till the end of the treatment.

2.5 Numerical Methods

For complex geometrical configurations, explicit solutions of the oxygen transfer problem, namely problem (14), are not available. Numerical simulations are the only way to apply the model to real cases. The discretization of these models, described in [30], is achieved by the finite element method. Because of the particular mathematical structure of the model, based on mixed-dimensional differential equations, no commercially available simulator can handle it. All simulations have been performed using an internal C++ code based on the GetFem++ open source library [34].

The main advantage of the mixed-dimensional formulation adopted here is that the discretizations of the equations defined in the tissue and in the vascular network are completely independent in terms of the computational grids and the numerical schemes. We discretize the vascular network branches as separate subdomains. Each of them is approximated by a piecewise straight line. The problems of blood flow and oxygen transfer are approximated using continuous piecewise-polynomial finite elements. The interstitial flow problem is approximated using mixed finite elements.

Because of the non-linear constitutive laws of the model, the problem formulated for each step of the sequence (14) has been further linearized by means of a fixed-point approach, and the solution of the coupled problem has been reached via an iterative process. At each iteration, the numerical discretization schemes described above provide a high-dimensional linear system of equations solved by means of state-of-the-art linear (iterative) solvers with suitable preconditioners. For further details on the computational methods, we refer to [23, 31].

The virtual tissue sample considered in this work consists of $500\,\mu\text{m} \times 500\,\mu\text{m} \times 500\,\mu\text{m}$ and has been discretized by means of a uniform tetrahedral mesh of 15 nodes on each side. Linear finite elements were used to perform the numerical discretization consisting of 4358 degrees of freedom (DOF). The discretization of the vascular network, instead, depends on the effective number of vessels introduced, precisely 28, 181, and 496 vessels, respectively, for the V2, V18, V36 (described below) corresponding to 560, 5021, and 9928 DOFs used for the discretization of each case. This numerical resolution has been considered satisfactory after a mesh-sensitivity analysis. In particular, the L^2-mean relative error in the approximation of oxygen concentration, C_t, between the considered mesh and a new lower resolution mesh (precisely $11 \times 11 \times 11$ points per side) was less than 1%.

3 Numerical Simulations of the Interaction of Radiotherapy and the Tumor Microenvironment

In this section, we investigate the role of the tumor microenvironment in radiotherapy using the mathematical and computational tools presented before. We start by reporting in Table 1 the set of parameters used to initialize the model (14), except for some specific aspects related to radiobiological data that will be handled case by case. The sensitivity of the model on these parameters has recently been studied in [45].

A set of Voronoi capillary networks was inherited from previous work [30, 31], allowing simulations of different capillary densities. In fact, the blood network is crucial in defining the oxygen distribution within the tumor, leading to hypoxic or normoxic regions [45]. More precisely, we adopted three types of vascular networks:

- V2: low capillary density characterized by 28 capillaries;
- V18: high capillary density made up of 181 capillaries;
- V36: hyper-vascularized, consisting of 496 capillaries.

Table 1 Set of parameters adopted to solve the oxygen transport model

Parameter	Definition	Value	Ref.
D	Characteristic length of the problem	$5 \cdot 10^{-4}$ m	[31]
R	Average radius of the capillary	$4 \cdot 10^{-6}$ m	[31]
K	Tissue hydraulic conductivity	$3.6 \cdot 10^{-17}$ m^2	[40]
μ_t	Interstitial fluid viscosity	$1.2 \cdot 10^{-3}$ cP	[31]
μ_v	Blood viscosity	$3 \cdot 10^{-3}$ cP	[31]
L_p	Wall hydraulic conductivity	$1.4 \cdot 10^{-10}$ m^2 s kg^{-1}	[35]
π_v	Oncotic plasma pressure	2639.34 Pa	[15]
π_t	Oncotic interstitial pressure	1999.5 Pa	[42]
σ	Reflection coefficient	0.82	[40]
σ_{oxy}	Oxygen reflection coefficient	0.0	–
P_L	Permeability coefficient	$3.5 \cdot 10^{-5}$ m/s	[7],[21]
D_v	Diffusivity coefficient in vessels	$2.18 \cdot 10^{-9}$ m^2/s	[24]
D_t	Diffusivity coefficient in tissue	$2.41 \cdot 10^{-9}$ m^2/s	[24]
V_{max}	Oxygen consumption rate in tissue	$2.47 \cdot 10^{-4}$ ml$_{O_2}$/cm^3/s	[47]
C_{in}	Vessel inlet oxygen concentration	$3 \cdot 10^{-3}$ ml$_{O_2}$/ml$_B$	
N	Bound coefficient	1.36 ml$_{O_2}$/g$_{Hb}$	
γ	Hill constant	2.64 –	
$MCHC$	Hb concentration in RBC	0.34 g$_{Hb}$/ml$_{RBC}$	
$P_{s_{50}}$	pO_2 at half saturation	27 mmHg	

For each of these networks, Murray's law was adopted to model the generation of the daughter vessel in the bifurcation process [31]. Furthermore, the ranges of capillary density considered were estimated based on the surface-to-volume ratio, namely the value of the lateral capillary surface over the volume of the tissue, precisely $S/V = 10.19,\ 70.90,\ 144.16\ \text{mm}^{-1}$ for cases V2, V18, V36 respectively.

We note that the virtual tissue sample of 500 μm × 500 μm × 500 μm captures a small portion of any reasonably developed vascularized tumor. In these conditions, the variability of network structure and influence of the artificial boundaries may play a non-negligible role, also depending on the quantity of interest. The sensitivity of oxygen transfer to vascular density and distribution has been extensively discussed in [45], showing that for oxygenation the physical parameters (that is, those in Table 1 dominate over the variability of the vascular network. More precisely, C_{in} and V_{max} are the factors that influence primarily the mean value of the partial pressure in the tissue. Based on this evidence, we claim that the results presented in this work would not be significantly affected if a larger tissue sample were used, with a considerable increase in computational cost. To overcome this limitation, we have recently developed a surrogate model of (8) that takes advantage of deep learning-accelerated model order reduction techniques, enabling to solve the equations of the model in fraction of the time needed by the current approach [46].

3.1 The Effect of Fractionated Radiotherapy

We study the effect of different fractionation patterns in order to maximize the effect of the total dose and simultaneously reduce the damage to normal tissue. New radiotherapy and online imaging technologies have allowed the spread of the so-called Stereotactic Body Radiation Therapy (SBRT) providing high doses (6–30 Gy) per fraction in a few fractions (1–8) [10, 27]. For this purpose, new ways of dose delivery have been proposed in recent years. A variety of dose patterns are generally reported in the literature, but there is a lack of agreement in the clinical radiobiology community on the optimum fractionation pattern. Thus, three radiotherapeutic patterns are considered here, from single-dose irradiation (1×48 Gy) to more realistic hypofractionated patterns (4×12 Gy, 6×8 Gy) [3].

We simulate poorly perfused tissue characterized by the V2 network. The hypofractionation patterns performed were 1×48 Gy, 4×12 Gy, and 6×8 Gy, displaying the same total dose ($D_{tot} = 48$ Gy) but released differently over time. Every fraction was delivered day by day, despite the large dose administered.

The main results of these simulations are shown in Fig. 1. The non-homogeneity depicted in SF is affected by the low vascular density assumed in this case. It is a possible representation of tumor tissue, where hypoxic regions coexist with normoxic ones to create a heterogeneous distribution of oxygen. We note that in the single-fraction treatment, SF reaches values close to the minimum ($SF \approx 10^{-3}$) near the vessels, indicated by the green spots on the map. However, it should be noted that 1×48 Gy is not a realistic treatment but a test case useful to make the comparison (30 Gy in 1 single fraction is the current "ceiling" used for very small tumors, [3]). For hypofractionated patterns, namely 4×12 Gy and 6×8 Gy, SF converges homogeneously to 10^{-3} at the end of the treatment (4° and 6° Fr, respectively) while the 1×48 Gy case is only limited to an average of 10^{-1}, despite the large dose delivered all at once. This demonstrates that if a tumor is characterized by low oxygen supply due to insufficient vascularization, as exemplified by the case of the V2 Voronoi network, single dose irradiation is less effective in determining cell death compared to fractionated treatments as it does not take advantage of reoxygenation after radiotherapy, as will be extensively discussed in the following section.

3.2 The Effect of Reoxygenation After Radiotherapy

In this section, we compare the effect of radiotherapy on a set of different vascular networks, to account for the heterogeneity of tissue targets. In particular, Voronoi networks V2, V18, and V36 were used to simulate tumors (more precisely metastases) that occur in various biological structures.

It is well known that hypoxic regions are more resistant to radiation therapy than normoxic regions [9, 14], reducing the effect of radiation damage. Thus,

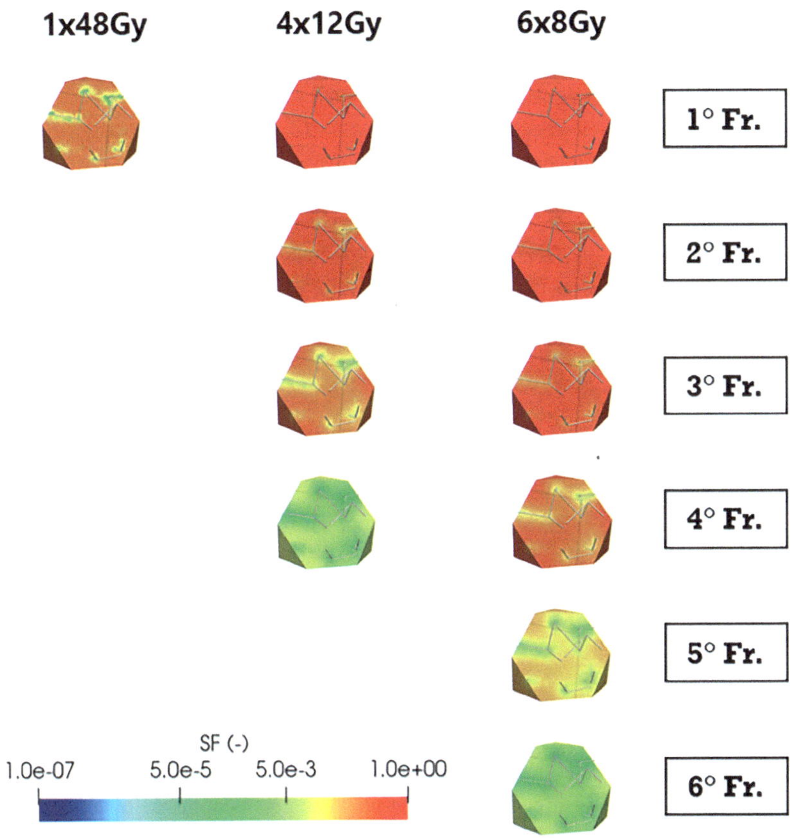

Fig. 1 Surviving fraction distribution within tumor volume. Different fractionation patterns are reported to test the efficacy of hypofractionated radiotherapy

tumor reoxygenation is one of the most important effects from a clinical point of view. It consists of improving tissue oxygenation in the tumor after each fraction of treatment. Hypoxic regions commonly arise in tumors, mainly due to the peculiar tumor microvasculature that originates during the chaotic process of tumor angiogenesis. In fact, in a tumor vessel network, oxygen distribution is affected by defects in the vascular structure, such as dead ends, tortuous paths, and leaky walls. As a consequence, some parts of the tumor are not properly perfused, leading to hypoxic conditions.

Table 2 reports the radiobiological data adopted for the tumor tissue. Here, Wenzl's parameters for the modified LQ model were adjusted to achieve the radiosensitivity properties of the lung, in particular, $\alpha = 0.16\ \text{Gy}^{-1}$ and $\alpha/\beta = 18\ \text{Gy}$ [19].

The relationship between oxygen distribution and the corresponding surviving fraction, intrinsically embedded in the radiobiological model, is depicted in Fig. 2.

Table 2 Radiobiological parameters for lung tumor

Parameter	Definition	Value	Unit of measure	Ref.
a_1	Wenzl's parameter	0.15950134	1/Gy	–
a_2	Wenzl's parameter	0.00174001	$\mu m/(Gy \cdot keV)$	–
a_3	Wenzl's parameter	0.03625030	1/Gy	–
a_4	Wenzl's parameter	0.00224752	$\mu m/(Gy \cdot keV)$	–
b_1	Wenzl's parameter	0.09654740	1/Gy	–
b_2	Wenzl's parameter	0.00362053	1/Gy	–
LET	Linear energy transfer	2	$keV/\mu m$	[48, 49]
K	Wenzl's O_2 concentration	2.5	mmHg	[48, 49]

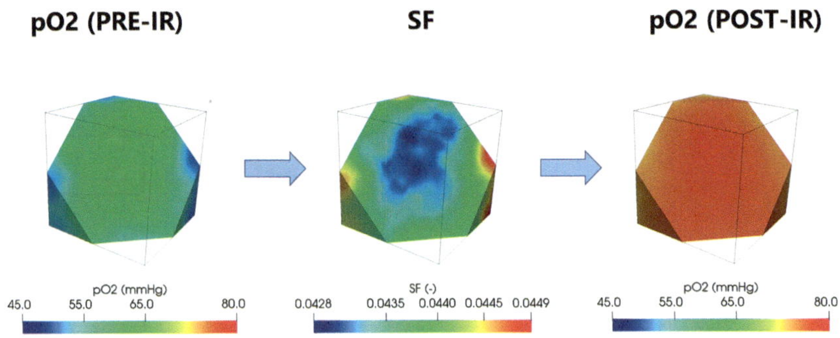

Fig. 2 Local effect of oxygen pressure distribution on the surviving fraction of tumor cells. Before irradiation (IR), the central portion of the tumor is well-oxygenated, leading to significant cell death, as demonstrated by the vast blue area in SF. Finally, the pO_2 resulting after irradiation is depicted, where a more homogeneous oxygen distribution is present thanks to the reduction of available tumor cells

The heterogeneity of pO_2 distribution is based on the capillary network adopted, that is, V36, in this case. We observe that the central portion of the cubic domain is characterized by a larger pO_2 (e.g. 65 mmHg) with respect to the periphery (e.g. 50 mmHg). As a consequence, the SF map resembles the oxygen one: portions characterized by high pO_2 are inevitably more sensitive to radiation exposure, causing significant cell death, where SF reaches values close to 0.0430, meaning that less than 5% of cells survived treatment in the central portion. After radiation exposure, the new cell distribution affects pO_2 in the next fraction, increasing the oxygen distribution in the central region, where a lower value of surviving cells was recorded. In addition, a more homogeneous oxygen pressure is generally achieved, reaching values around 76 mmHg.

In Fig. 3, we compare the effect of oxygenation on radiotherapy in three biological tissues with different vascular densities. Here, the same radiation pattern is provided in all the cases depicted, namely, a 4×12 Gy, with the conventional one fraction per day, despite the high dose delivered.

A ripple effect is induced when the oxygen pressure varies within the treatment: the SF calculated for the previous fraction interacts with the oxygen consumption

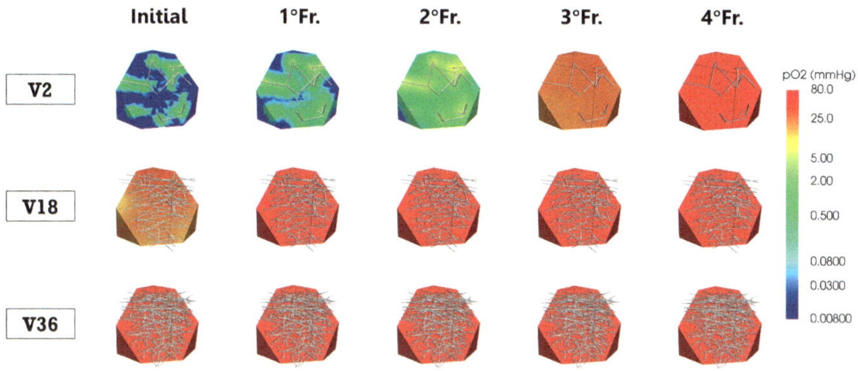

Fig. 3 Evolution of oxygen partial pressure in tumor tissue according to the phase of radiotherapy cycle. A comparison among three types of vascular network is reported to emphasize the effect of reoxygenation with different initial oxygen distribution

Table 3 Final value of SF in a 4×12 Gy fractionation pattern considering the effect of the vascular network. A lower value of SF is observed for well-oxygenated conditions, namely V18 and V36

	V2	V18	V36
SF_{Final}	$4.85 \cdot 10^{-3}$	$2.04 \cdot 10^{-5}$	$1.71 \cdot 10^{-5}$

coefficient, modifying the amount of oxygen consumed by the cells. In this way, the consumption rate is reduced, leading to higher oxygen levels and a more uniform oxygen distribution in the tissue after radiotherapy.

An example of effective reoxygenation is reported in the case of low vascular density, namely V2, where tumor reoxygenation is crucial to achieving tumor control by radiotherapy. Initially, the oxygen distribution is poor in the tumor, limiting radiation damage. As the treatment progresses, radiotherapy reduces the number of surviving cells and thus prepares the environment for the next fraction. Therefore, oxygen could spread homogeneously within the tumor, reaching locations far from vessels. At the end of treatment, the tumor reaches a fully oxygenated state, with a mean value of 71.8 mmHg. In the V2 case, the effect of the initial distribution of oxygen is evident, especially in affecting inter-fractions variations, unlike what happens for V18 and V36, where a desired oxygen distribution is achieved after only one fraction. Indeed, the high vascular density of these two latter cases generates a uniform oxygen distribution from the beginning of treatment. In other words, radiotherapy is more effective if the tumor is fully oxygenated, as illustrated in Table 3.

Another interesting aspect relies on the comparison of the oxygen distribution using different fractionation patterns on the same network (V2, in this case). The results confirmed that fractionated patterns improve oxygen distribution; see Fig. 4. Indeed, the mean value of the partial oxygen pressure is 71.8 mmHg in the 6×8 Gy pattern (blue line in the figure) and 71.0 mmHg in the case 4×12 Gy (red). The

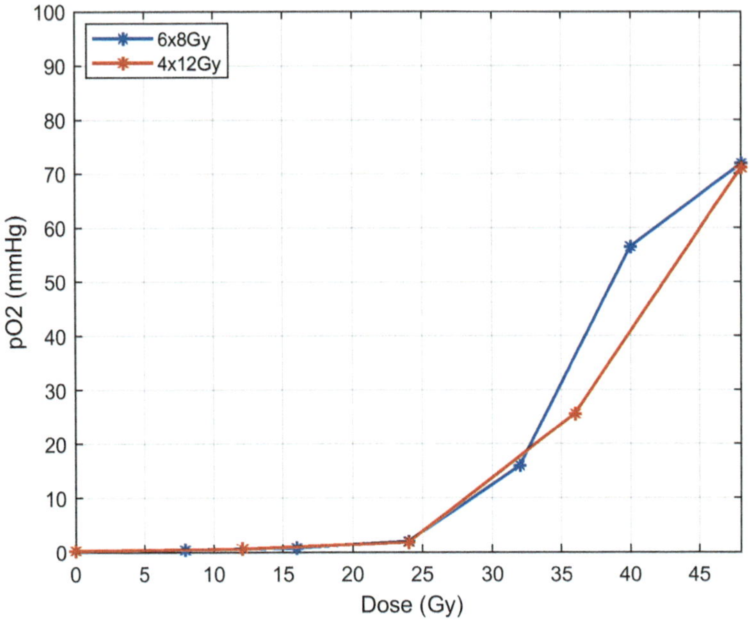

Fig. 4 Comparison of pO_2 in 4×12 Gy and 6×8 Gy in lung tumors. Low-density vascular network, V2, was adopted to emphasize the differences in reoxygenation

path to reach the final point is different in the two cases, with the more fractionated therapy guaranteeing higher oxygenation throughout the treatment. Providing a low dose per day results in a less aggressive treatment that allows cell renewal and remodeling of the vasculature through which oxygen diffusion is enhanced. A slight difference is obtained between the two patterns in this scenario, since vasculature remodeling was not included. Since having large dose fractions, such as 12 Gy, may cause serious vascular damage, we expect a greater contribution of blood vessels in reoxygenation as the vascular response to irradiation is introduced. As a result, oxygen transport would be impaired.

3.3 The Effect of Vascular Modifications Due to Radiotherapy

The impact of radiation on the microvasculature of a tumor is dose-dependent. High doses above 10 Gy can completely compromise the integrity of the vessel. Damage to the microvascular network can increase the number of hypoxic regions within the tumor. In these regions, the effectiveness of radiation therapy is further limited. The main effects of irradiation on blood vessels are mainly vascular collapse, degeneration of the vascular wall, and inflammation, increasing vascular leakage and permeability. Here, we focus mostly on the second factor, quantified as the

Table 4 Permeability variation following radiation exposure

Dose	Cells irradiated	ΔP_L	Ref.
2 Gy	Primary HUVEC	+25%	[22]
4 Gy	Primary HUVEC	+35%	[22]
6×4.5 Gy	Imaging (NSCLC patients)	+8.4% (2°fr.), +44.8% (4°fr.), +20.5% (6°fr.)	[28]

increase in vascular permeability due to radiation therapy. The role of permeability is critical in defining the release of oxygen from vessels to tissue, as it determines whether substances can cross the endothelial barrier or not. In fact, a change in permeability has a pivotal effect, since a variation in oxygen concentration in the tissue greatly influences the effect of radiotherapy and the final clinical outcome.

A huge discrepancy has been observed in the variation of permeability following radiotherapy. Tumor heterogeneity is the main cause of this dispersion in the data. The size of the tumor, the type of cells irradiated, and the stage of the tumor are all elements that influence the surrounding microvasculature and thus the general response to radiation. In particular, a general trend for increased vessel permeability is described in the literature immediately after irradiation. Table 4 provides an example of the variability of the data.

Based on the aforementioned data, we assumed a permeability increase of +35% following a photon-based single dose of 4 Gy, as reported in [22]. In particular, it was assumed that permeability would increase after each delivery. In addition, a dose-dependent behavior was included, as suggested in many works [2, 6, 16, 20, 29, 50]. We assumed a linear increase in the relationship between vessel permeability and the dose administered, given that a model for such a dose relationship is not defined.

The increase in permeability was simulated in a breast cancer environment, assuming $\alpha = 0.3\ \text{Gy}^{-1}$ and $\alpha/\beta = 3$ Gy. The total treatment time was 12 days, with the usual 5 fr/week pattern. The oxygen parameters were the same as previously used (see Table 1), with an initial value of vessel permeability $P_L = 3.5 \cdot 10^{-5}$ m/s. As an initial condition, we obtained the oxygen content provided by a V2 Voronoi network characterized by significant heterogeneity. The tumor was then irradiated with a dose fraction of 4 Gy, inducing a increase in permeability of +35%. Moreover, a control case is illustrated, where the permeability was assumed constant throughout the treatment. Assuming a +35% increment, the value of permeability rises to $4.725 \cdot 10^{-5}$ m/s after the first fraction, leading to a local increase in the partial pressure of oxygen in the region surrounding the vessels (Fig. 5, 1 Fr.). Compared to the related control case, the higher partial oxygen pressure (green area) spreads out in the high-permeability case, achieving a more homogeneous oxygen distribution. This effect is emphasized during treatment, as shown in the 3° fraction, when the permeability (P_L) was $8.611 \cdot 10^{-5}$ m/s, confirming the trend previously described. More precisely, a mean pO_2 around 15 mmHg is achieved when assuming variation in permeability, while only 2–5 mmHg are present in the

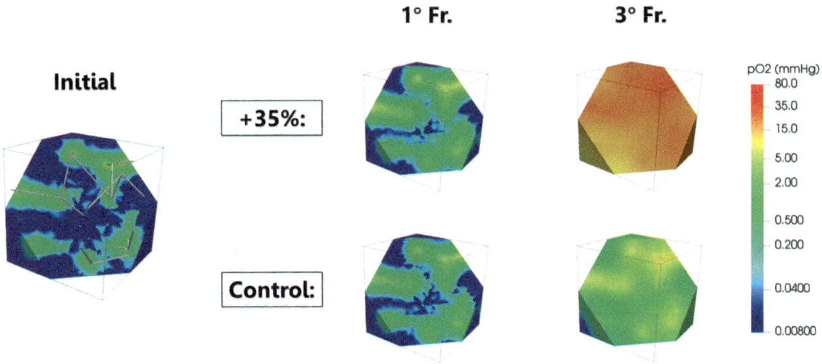

Fig. 5 Effect of permeability variation in consecutive fractions. A comparison with a control case (constant permeability) is provided to appreciate the different oxygen distribution related to the permeability increment

control case (Fig. 5, 3°Fr.). The results demonstrate that the permeability increase during treatment generates better-oxygenated area all over the domain.

The results described above are the consequence of the positive superposition of the two effects, increased permeability, and decreased consumption. In fact, tissue oxygenation is facilitated by a decrease in the oxygen consumption rate, which is related to the death of tumor cells in the tissue. More precisely, greater radiation damage is induced in these regions because of the high oxygen content, leading to a lower surviving fraction. Consequently, treatment is more effective when permeability damage is considered, as demonstrated by the final value of SF we obtained ($SF_{control} = 1.34 \cdot 10^{-8}$; $SF_{\Delta P_L} = 1.65 \cdot 10^{-9}$). In particular, a lower value of 87% is recorded in SF when an increase in permeability 35% is considered.

Finally, the variation in oxygen pressure throughout treatment is investigated in Fig. 6. As usual, the mean value of pO_2 in the tumor tissue is reported in each fraction. Assuming an increase in vessel permeability, a higher value of mean oxygen pressure is expected within a few fractions. Therefore, in the case of +35% and +52.5% increment, the pO_2 curve shifts towards the left, reaching the maximum level of oxygenation before the control curve (constant permeability). This confirms the previously shown results, where oxygenation is enhanced because of increased permeability. However, as soon as the maximum is reached, the +35% and +52.5% curves suddenly decrease, despite the constant increase in permeability. This peculiar phenomenon can be explained from a mathematical point of view. In fact, equilibrium is reached in terms of oxygen concentration between blood vessels (C_v) and tumor tissue (C_t) as permeability increases since oxygen diffuses from the higher to the lower concentrated districts until a concentration gradient is available. As a consequence, the term $(C_v - C_t)$ of Eq. (8), which involves oxygen transport through the vessel barrier, is rapidly reduced during the fractionation process, affecting the variation of oxygen concentration (C_t) in the following fraction. In other words, the constant increase in permeability is balanced by a

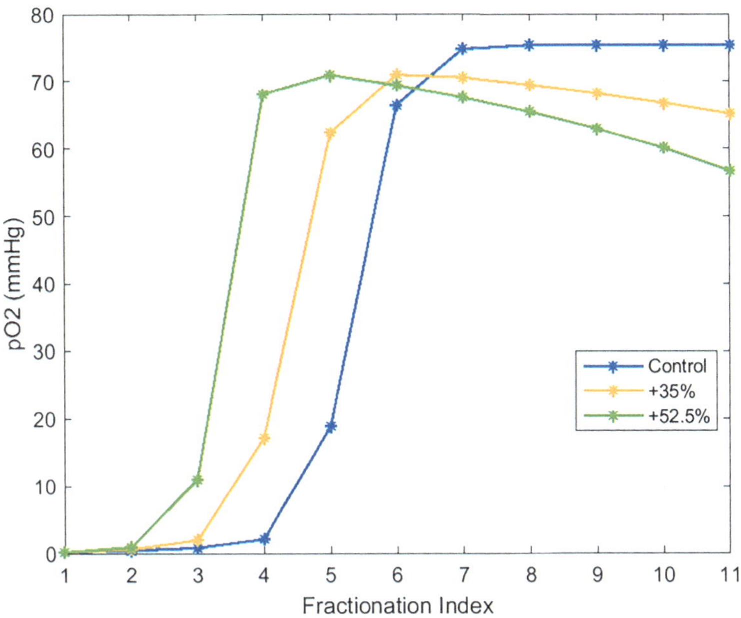

Fig. 6 Mean oxygen pressure for different fractions of the treatment. Increasing the dose per fraction has a significant effect on nominal permeability variation, resulting in an enhancement of the oxygenation process. Indeed, the maximum point is reached in fewer fractions in the +52.5% case (green curve) as compared to the +35% one (yellow curve)

decrease in the transmural concentration gradient. In this way, the oxygen support to the tissue decreases, while the oxygen consumption rate remains unchanged. The combination of these effects explains the slight decrease in oxygen levels in the case of highly permeable vessels. Instead, for the control curve, no significant variations are reported. The constant permeability maintains equilibrium without involving any variation in C_t.

4 Discussion and Conclusions

In this work, we study the effect of oxygen concentration in radiotherapy through the coupling of a radiobiological model with an oxygen transport model in the vascular microenvironment. First, the Linear-Quadratic radiobiological model was modified to include the dependence on the partial pressure of oxygen, as proposed by Wenzl et al. [48, 49]. This allows us to relate the surviving fraction with the amount of oxygen available through the tumor radiobiological properties α and β. Second, the radiobiological model was incorporated into a sophisticated oxygen transport model, where the oxygen exchange in the microvascular environment was simulated. A numerical approach was adopted to solve the complex system of equations resulting from the above-mentioned coupling.

The proposed model was particularly useful for the analysis of the phenomenon of reoxygenation in the tumor microenvironment. During treatment, radiation therapy progressively reduced the number of viable tumor cells, affecting oxygen consumption in the tumor. As a result, an increase in oxygenation is observed, assuming a constant blood washout between consecutive fractions. As a limitation of the model, tumor reoxygenation may also include other aspects related to the tumor microenvironment, such as vascular remodeling. Rearrangement of the microvascular network after radiotherapy greatly affects oxygen redistribution and should therefore be included in this model.

In terms of the study of the permeability variation, instead, we remark on the important simplifications adopted to model the dose-dependent behavior. The results show that the increase in permeability enhances reoxygenation, but we assume a linear relationship is assumed between the increase in permeability and the dose per fraction delivered. More research is required to model the dose-dependent effect on permeability. Also, due to vascular damage, an abnormal increase in permeability is probably related to a rise in interstitial fluid pressure and undesired extravasation of large molecules (advective phenomena), causing vessel collapse and thus decreasing blood vessel density in the tumor.

Simulated treatments show, on average, good efficacy in tumor eradication, quantified by the low value achieved in the surviving fraction (SF). Compared to clinical observations, these results look quite optimistic. Tumor recurrence remains a problem, despite the simulations performed and the related promising results. This discrepancy between simulations and reality may be explained by observing that we have modeled a well-oxygenated tumor environment characterized by a homogeneous distribution of oxygen partial pressure that explains such optimistic results in survival fraction after radiotherapy. In fact, a limitation of this work consists in considering a rather small sample describing a well-perfused microenvironment characterized by a regular and organized vascular network, generally not observed in tumors. This limitation was partially overcome by simulating vascular networks with low vascular density. As expected, hypofractionated patterns showed low performance in this case due to the poor reoxygenation that induces a hyperradioresistant response.

In addition to that, this work supports the importance of oxygenation and its changes in the framework of radiation therapy, and the proposed model can be used to analyze the tumor microenvironment during fractionated radiotherapy.

Acknowledgments Paolo Zunino acknowledges the support of the grant MUR PRIN 2022 No. 2022WKWZA8 *Immersed methods for multiscale and multiphysics problems* (IMMEDIATE). The present research is part of the activities of the Dipartimento di Eccellenza 2023-2027 project, funded by MUR. Piermario Vitullo and Paolo Zunino are members of the Gruppo Nazionale per il Calcolo Scientifico (GNCS), Istituto Nazionale di Alta Matematica (INdAM).

References

1. Baker, D.G., Krochak, R.J.: The response of the microvascular system to radiation: a review. Cancer Investig. **7**(3), 287–294 (1989). https://doi.org/10.3109/07357908909039849
2. Barker, H.E., Paget, J.T.E., Khan, A.A., Harrington, K.J.: The tumour microenvironment after radiotherapy: mechanisms of resistance and recurrence. Nat. Rev. Cancer **15**, 409–425 (2015)
3. Bedard, G., McDonald, R., Poon, I., Erler, D., Soliman, H., Cheung, P., Chung, H., Chu, W., Loblaw, A., Chow, E., Sahgal, A.: Stereotactic body radiation therapy for non-spine bone metastases–a review of the literature. Ann. Palliat. Med. **5**, 58–66 (2016). https://doi.org/10.3978/j.issn.2224-5820.2015.07.01
4. Brenner, D.J., Hlatky, L.R., Hahnfeldt, P.J., Huang, Y., Sachs, R.K.: The linear-quadratic model and most other common radiobiological models result in similar predictions of time-dose relationships. Radiat. Res. **150**, 83–91 (1998)
5. Buffa, F.M.: Fundamental radiobiology and its application to radiation oncology. In: Lemoigne, Y., Caner, A. (eds.) Radiotherapy and Brachytherapy, pp. 3–9. Springer Netherlands, Dordrecht (2009)
6. Castle, K.D., Kirsch, D.G.: Establishing the impact of vascular damage on tumor response to high-dose radiation therapy. Cancer Res. **79**, 5685–5692 (2019)
7. Cattaneo, L., Zunino, P.: A computational model of drug delivery through microcirculation to compare different tumor treatments. Int. J. Numer. Method Biomed. Eng. **30**(11), 1347–1371 (2014)
8. De Ruysscher, D., Niedermann, G., Burnet, N., Siva, S., Lee, A., Hegi-Johnson, F.: Radiotherapy toxicity. Nat. Rev. Dis. Primers **5**(1), 13 (2019)
9. Eui Jung Moon, K.P., Olcina, M.M.: The importance of hypoxia in radiotherapy for the immune response, metastatic potential and FLASH-RT. Int. J. Radiat. Biol. **98**(3), 439–451 (2022). https://doi.org/10.1080/09553002.2021.1988178
10. Grimm, J., Marks, L.B., Jackson, A., Kavanagh, B.D., Xue, J., Yorke, E.: High dose per fraction, hypofractionated treatment effects in the clinic (HyTEC): an overview. Int. J. Radiat. Oncol. Biol. Phys. **110**, 1–10 (2021). https://doi.org/10.1016/j.ijrobp.2020.10.039
11. Grogan, J.A., Markelc, B., Connor, A.J., Muschel, R.J., Pitt-Francis, J.M., Maini, P.K., Byrne, H.M.: Predicting the influence of microvascular structure on tumor response to radiotherapy. IEEE Trans. Biomed. Eng. **64**, 504–511 (2017)
12. Hanna, T.P., King, W.D., Thibodeau, S., Jalink, M., Paulin, G.A., Harvey-Jones, E., O'Sullivan, D.E., Booth, C.M., Sullivan, R., Aggarwal, A.: Mortality due to cancer treatment delay: systematic review and meta-analysis. BMJ **371** (2020). https://doi.org/10.1136/bmj.m4087
13. Hartung, G., Badr, S., Moeini, M., Lesage, F., Kleinfeld, D., Alaraj, A., Linninger, A.: Voxelized simulation of cerebral oxygen perfusion elucidates hypoxia in aged mouse cortex. PLoS Comput. Biol. **17**, e1008584 (2021)
14. Horsman, M.R., Mortensen, L.S., Petersen, J.B., Busk, M., Overgaard, J.: Imaging hypoxia to improve radiotherapy outcome. Nat. Rev. Clin. Oncol. **9**, 674–687 (2012)
15. Jain, R.K., Tong, R.T., Munn, L.L.: Effect of vascular normalization by antiangiogenic therapy on interstitial hypertension, peritumor edema, and lymphatic metastasis: insights from a mathematical model. Cancer Res. **67**, 2729–2735 (2007)
16. Jarosz-Biej, M., Smolarczyk, R., Cichoń, T., Kułach, N.: Tumor microenvironment as a "game changer" in cancer radiotherapy. Int. J. Mol. Sci. **20**, 3212 (2019)
17. Jarzyńska, M.: The application of practical Kedem-Katchalsky equations in membrane transport. Open Phys. **4**, 429–438 (2006)
18. Kellerer, A.M., Rossi, H.H.: RBE and the primary mechanism of radiation action. Radiat. Res. **47**, 15–34 (1971)
19. Klement, R.J.: Radiobiological parameters of liver and lung metastases derived from tumor control data of 3719 metastases. Radiother. Oncol. **123**, 218–226 (2017)

20. Kocher, M., Treuer, H., Voges, J., Hoevels, M., Sturm, V., Müller, R.P.: Computer simulation of cytotoxic and vascular effects of radiosurgery in solid and necrotic brain metastases. Radiother. Oncol. **54**, 149–156 (2000)
21. Köppl, T., Vidotto, E., Wohlmuth, B.: A 3D-1D coupled blood flow and oxygen transport model to generate microvascular networks. Int. J. Numer. Methods Biomed. Eng. **36**, e3386 (2020)
22. Kouam, P.N., Rezniczek, G.A., Adamietz, I.A., Bühler, H.: Ionizing radiation increases the endothelial permeability and the transendothelial migration of tumor cells through adam10-activation and subsequent degradation of ve-cadherin. BMC Cancer **19**, 958 (2019)
23. Laurino, F., Zunino, P.: Derivation and analysis of coupled PDEs on manifolds with high dimensionality gap arising from topological model reduction. ESAIM: M2AN **53**(6), 2047–2080 (2019)
24. Lücker, A., Weber, B., Jenny, P.: A dynamic model of oxygen transport from capillaries to tissue with moving red blood cells. Am. J. Physiol. Heart Circ. Physiol. **308**, H206–H216 (2015)
25. McMahon, S.: The linear quadratic model: usage, interpretation and challenges. Phys. Med. Biol. **64**(1) (2019)
26. McMahon, S.J., Prise, K.M.: Mechanistic modelling of radiation responses. Cancers **11**, 205 (2019)
27. Navarria, P., Baldaccini, D., Clerici, E., Marini, B., Cozzi, L., Franceschini, D., Bertuzzi, A.F., Quagliuolo, V., Torri, V., Colombo, P., Franzese, C., Bellu, L., Scorsetti, M.: Stereotactic body radiation therapy for lung metastases from sarcoma in oligometastatic patients: a phase 2 study. Int. J. Radiat. Oncol. Biol. Phys. **114**, 762–770 (2022). https://doi.org/10.1016/j.ijrobp.2022.08.028
28. Ng, Q.S., Goh, V., Milner, J., Padhani, A.R., Saunders, M.I., Hoskin, P.J.: Acute tumor vascular effects following fractionated radiotherapy in human lung cancer: in vivo whole tumor assessment using volumetric perfusion computed tomography. Int. J. Radiat. Oncol. Biol. Phys. **67**, 417–424 (2007)
29. Park, H.J., Griffin, R.J., Hui, S., Levitt, S.H., Song, C.W.: Radiation-induced vascular damage in tumors: implications of vascular damage in ablative hypofractionated radiotherapy (SBRT and SRS). Radiat. Res. **177**, 311–327 (2012)
30. Possenti, L., Casagrande, G., Di Gregorio, S., Zunino, P., Costantino, M.: Numerical simulations of the microvascular fluid balance with a non-linear model of the lymphatic system. Microvasc. Res. **122**, 101–110 (2019)
31. Possenti, L., di Gregorio, S., Gerosa, F., Raimondi, G., Casagrande, G., Costantino, M., Zunino, P.: A computational model for microcirculation including Fahraeus-Lindqvist effect, plasma skimming and fluid exchange with the tissue interstitium. Int. J. Numer. Methods Biomed. Eng. **35**(3), e3165 (2019)
32. Possenti, L., Cicchetti, A., Rosati, R., Cerroni, D., Costantino, M.L., Rancati, T., Zunino, P.: A mesoscale computational model for microvascular oxygen transfer. Ann. Biomed. Eng. **49**, 3356–3373 (2021). https://doi.org/10.1007/s10439-021-02807-x
33. Ratosa, I., Jenko, A., Oblak, I.: Breast size impact on adjuvant radiotherapy adverse effects and dose parameters in treatment planning. Radiol. Oncol. **52**(3), 233–244 (2018). https://doi.org/10.2478/raon-2018-0026
34. Renard, Y., Poulios, K.: GetFEM: automated fe modeling of multiphysics problems based on a genericweak form language. ACM Trans. Math. Softw. **47**(1), 1–31 (2021)
35. Rippe, B., Kamiya, A., Folkow, B.: Simultaneous measurements of capillary diffusion and filtration exchange during shifts in filtration-absorption and at graded alterations in the capillary permeability surface area product (PS). Acta Physiol. Scand. **104**, 318–336 (1978)
36. Rosati, R.: A multi-scale computational model for micro-vascular oxygen transfer applied to radiotherapy. Master thesis. Politecnico di milano. (2019). https://hdl.handle.net/10589/153539
37. Santiago, A., Barczyk, S., Jelen, U., Engenhart-Cabillic, R., Wittig, A.: Challenges in radiobiological modeling: can we decide between LQ and LQ-L models based on reviewed clinical NSCLC treatment outcome data? Radiat. Oncol. **11**, 67 (2016)

38. Scott, J.G., Fletcher, A.G., Anderson, A.R.A., Maini, P.K.: Spatial metrics of tumour vascular organisation predict radiation efficacy in a computational model. PLoS Comput. Biol. **12**, e1004712 (2016)
39. Secomb, T., Hsu, R., Park, E., Dewhirst, M.: Green's function methods for analysis of oxygen delivery to tissue by microvascular networks. Ann. Biomed. Eng. **32**(11), 1519–1529 (2004)
40. Sefidgar, M., Soltani, M., Raahemifar, K., Sadeghi, M., Bazmara, H., Bazargan, M., Naeenian, M.M.: Numerical modeling of drug delivery in a dynamic solid tumor microvasculature. Microvasc. Res. **99**, 43–56 (2015)
41. Sinclair, W.: The shape of radiation survival curves of mammalian cells cultured in vitro. International atomic energy agency. Panel on biophysical aspects of radiation quality. Vienna (Austria). p. 21–43 (1966)
42. Sweeney, P., D'esposito, A., Walker-Samuel, S., Shipley, R.: Modelling the transport of fluid through heterogeneous, whole tumours in silico. PLoS Comput. Biol. **15**(6), e1006751 (2019)
43. van Leeuwen, C.M., Oei, A.L., Crezee, J., Bel, A., Franken, N.A.P., Stalpers, L.J.A., Kok, H.P.: The alfa and beta of tumours: a review of parameters of the linear-quadratic model, derived from clinical radiotherapy studies. Radiat. Oncol. **13**(1), 96 (2018). https://doi.org/10.1186/s13014-018-1040-z
44. Venkatesulu, B.P., Mahadevan, L.S., Aliru, M.L., Yang, X., Bodd, M.H., Singh, P.K., Yusuf, S.W., ichi Abe, J., Krishnan, S.: Radiation-induced endothelial vascular injury: a review of possible mechanisms. JACC: Basic Transl. Sci. **3**(4), 563–572 (2018). https://doi.org/10.1016/j.jacbts.2018.01.014
45. Vitullo, P., Cicci, L., Possenti, L., Coclite, A., Costantino, M.L., Zunino, P.: Sensitivity analysis of a multi-physics model for the vascular microenvironment. Int. J. Numer. Methods Biomed. Eng., e3752 (2023). https://doi.org/10.1002/cnm.3752
46. Vitullo, P., Colombo, A., Franco, N.R., Manzoni, A., Zunino, P.: Nonlinear model order reduction for problems with microstructure using mesh informed neural networks. Finite Elem. Anal. Des. **229**, 104068, (2024) ISSN 0168-874X. https://doi.org/10.1016/j.finel.2023.104068
47. Welter, M., Fredrich, T., Rinneberg, H., Rieger, H.: Computational model for tumor oxygenation applied to clinical data on breast tumor hemoglobin concentrations suggests vascular dilatation and compression. PLoS One **11**, e0161267 (2016)
48. Wenzl, T., Wilkens, J.J.: Modelling of the oxygen enhancement ratio for ion beam radiation therapy. Phys. Med. Biol. **56**, 3251–3268 (2011)
49. Wenzl, T., Wilkens, J.J.: Theoretical analysis of the dose dependence of the oxygen enhancement ratio and its relevance for clinical applications. Radiat. Oncol. **6**, 171 (2011)
50. Wijerathne, H., Langston, J.C., Yang, Q., Sun, S., Miyamoto, C., Kilpatrick, L.E., Kiani, M.F.: Mechanisms of radiation-induced endothelium damage: emerging models and technologies. Radiother. Oncol. **158**, 21–32 (2021)

Microvascular Modeling for Medical Imaging and Toxicity Assessment

Jérôme Kowalski, Dirk Drasdo, Peter Kottman, Jules Dichamp, and Irene Vignon-Clementel

Abstract Dynamic-contrast-enhanced medical imaging is a non-invasive method for detecting tissue abnormalities like tumors, evaluating organ function, or assessing drug efficacy. Tracers injected into the body reflect tissue perfusion and function. Characterizing this 'reflection' *in-vivo* is challenging: *in-silico* methods are being developed for better image interpretation.

Moreover in organs such as the liver, cells are nourished by networks of small vessels. This vasculature is prone to developing diseases due to the liver detoxification function. Understanding drug effects on the liver involves invasive *in-vivo* toxicity experiments, reducing subject lifespan. *In-vitro* experiments are simpler; yet their relevance needs to be assessed. *in-silico* models are increasingly invoked to link the two experiment types, reducing the need for *in-vivo* toxicity tests.

This chapter contains three *in-silico* models that, based on a common mathematical framework, simulate blood flow and drug transport in and out of a diseased micro-vascular network, and their effect on cells. To that end, the main pillars of multi-phase multi-species flow and transport models are first derived from primary principles and then transcribed for a vascularized tissue. The chapter provides overall a methodology to link changes in vascular network architectures, micro-circulation consequences and some macroscale biomarkers relevant for dynamic medical imaging or toxicity experiments.

Keywords Microcirculation · Mathematical model · Blood flow · Transport · Tumor · Liver · Disease · Vasculature · Toxicity · Dynamic medical imaging

---

J. Kowalski · P. Kottman · J. Dichamp · I. Vignon-Clementel (✉)
INRIA, Saclay Ile-de-France Research Center, Paris, France
e-mail: jerome.kowalski@inria.fr; peter.kottman@inria.fr; jules.dichamp@inria.fr; irene.vignon-clementel@inria.fr

D. Drasdo
INRIA, Saclay Ile-de-France Research Center, Paris, France

IfADo - Leibniz Research Centre for Working Environment and Human Factors, Dortmund, Germany
e-mail: dirk.drasdo@inria.fr

A. Linninger et al. (eds.), *Quantitative Approaches to Microcirculation*,
SEMA SIMAI Springer Series 36, https://doi.org/10.1007/978-3-031-58519-7_3

1 Introduction

Modelling flow and transport inside living organisms at all scales has always been a subject of great interest (see the work of scientists such as Harvey, Bernoulli, Poiseuille, …). While the appeal of modelling macroscopic flow is quite natural, due to relative ease of comparison of model results to *in-vivo* measurements in large vessels [1], recent advances in imaging techniques have provided an argument to further model flow and transport inside organ micro-structure [21]. These models possess the capacity to subsequently validate simplified representations commonly employed in medical imaging, such as larger-scale kinetic models. Alternatively, they can directly assist in pathology assessment and therapy optimization since many disease processes first develop at the functional unit of an organ, hence at the microstructure level.

The liver is composed of about a million of smallest repetitive anatomical and physiological units, called liver lobules. They are approximately prism-shaped with an hexagonal basis and perfused by complex capillary networks, connecting branches of the feeding portal veins and hepatic arteries localized at the lobule borders. After having travelled through the capillaries, called sinusoids in the liver, blood leaves the lobule through the central vein branch located in the lobule center. Essential nutrients, harmful toxins, and pharmaceutical compounds that ingress the liver traverse these intricate vascular networks. Along their course, these substances undergo absorption by hepatocytes, the specialized cellular components of the liver responsible for their metabolic processing. Metabolic products are either secreted back into the capillaries, or into the biliary network from where they are carried towards the gallbladder. The hepatocyte uptake rates and metabolic processes vary across different lobule regions—phenomenon called zonation—as well as across different parts of the liver. Modifications in one or more of any of these processes can lead to pathologies, ranging from mild to severe, and even fatal.

As a more general example of this, tumor evolution can be assessed through various histopathological features, including cell-specific fingerprints (such as atypical nuclei, an atypic nucleus-plasma-relation, ...), tissue specific fingerprints (such as higher cell density, invasion of stroma, ductual reaction) or, for example a remodelling of microvasculature by either change of fenestration, increase, loss of capillaries or micro-architectural changes. However, reliable spatial information about these features can hardly be obtained non-invasively, and require biopsies or tissue samples. As biopsies can lead to serious adverse effects, they disqualify for extensive longitudinal data collection, which is why non-invasive imaging modalities are gaining increasing importance in attempting to infer microarchitectural information. This is where imaging techniques such as diffusion-weighted magnetic resonance imaging (DW-MRI) or dynamic contrast-enhanced perfusion imaging (DCE-PI: DCE-MRI, DCE-CT, DCE-US, ...) become prominent, providing an indirect mesoscopic view of tissue regions of interest (ROI). The interpretation of the signal intensities in both DW-MRI and DCE-PI is far from trivial. DW-MRI signal intensity has been shown in lung cancer to correlate inversely-linear with the microscopic cell densities by comparison of DW-MRI signals to histological investigations of the same tissue samples [22, 23]. Signal from DCE-PI images is related to information at histological level via tracer kinetics (TK) models. Although there are several TK models of varying complexity, their validation is not straightforward. Here, microcirculation models can serve to provide *in silico* validation of TK multi-compartment models and thus improve perspective methods for non-invasive tissue pathology classification [21].

Modelling blood flow in microvascular networks presents several challenges. First, due to the small vessel size relative to the erythrocytes (red blood cells), blood at a scale of small vessel diameters cannot be considered as a simple Newtonian fluid such that standard continuum blood flow models do not apply anymore. The presence of cells transported with the blood reduces the flow speed in vessels with a diameter between about 20 and 200 micro-meters, called the Fåhræus-Lindqvist effect, by organising cells within chains close to the blood vessel axis thereby generating a cell-depleted boundary layer. Below a diameter of about 20 micrometers, the erythrocytes start to touch the walls again hence the flow speed decreases again. An elegant way to account for these effects is by an empirical formula for blood viscosity (Pries and Secomb, [17]), which replaces the constant viscosity parameter in the standard continuous description for Newtonian fluids. Another phenomena due to blood heterogeneity is plasma skimming, connected to the uneven distribution of erythrocytes at network bifurcations [2].

Other than being an important field of study on its own, the flow in tissue micro-architecture constitutes a necessary parameter to model transport in the blood. In the liver, this determines how metabolites and toxins are transported into and out of the liver thereby contributing to liver function. A solid understanding of these processes and their modifications in liver disease would allow for better—and possibly patient-specific—prognostics and therapy planning. However, due to the complexity of vessel networks and metabolic processes involved, modelling such a system at its full detail is currently out of reach, thus calling for model reduction techniques [2, 5]. These have to capture the impact on the model of changes in

- network geometry,
- vessel radii,
- cell uptake kinetics,
- concentration input as a function of time,

among others. Simultaneously, such spatio-temporal models can serve as a *virtual* validation of multi-compartment TK models: one can simulate tracer transport in these models and study in how far a certain TK model, usually highly simplified due to lumping many parameters together, is able to infer characteristic quantitative tissue parameters. This may enable to infer micro-architectural information on tissue in disease, for example, to detect the presence, the micro-architecture, and pathophysiology of tumors, or assess their treatment. An example of such validation workflow consists in [21]:

1. constructing a representative *in-silico* vascularized tissue,
2. modelling flow and transport across a tissue to create *in-silico* dynamic contrast-enhanced perfusion images,
3. extracting representative quantities such as the contrast agent concentration evolution in time or its mean transit time,
4. interpreting the representative quantities via a chosen TK model,
5. inverse fitting the TK model back onto the constructed vascularized tissue,
6. comparing the resulting parameter maps with the originally computed *in-silico* values.

The focus of this chapter is on methods and results of microcirculation modelling relevant to applications outlined above, illustrating how they can provide clinically relevant information about the system in question. Although these models present differences and were developed independently, we realized that in fact they emerge from a common mathematical framework. Therefore the following methods section starts by establishing this baseline. The foundational principles underlying multi-phase multi-species mass transport models in the context of micro-architectural organ tissue are derived step-by-step from fundamental axioms. The goal is to explain the main ideas behind such derivation.

The results are dedicated to three applications. First, we present an *in-silico* model of tumor-invaded tissue perfusion, taking into account blood flow and transport of an injected compound in vessels as well as its diffusion in the interstitial space. We show how representative biomarkers such as mean transit time can provide tumor characteristics with the help of perfusion images. We then present two models of liver lobule perfusion: first focusing on microcirculation of blood and an injected compound in the vascular network; and then adding exchanges with surrounding hepatocytes and model APAP (a.k.a. paracetamol) detoxification as an example of liver function. We show how modification in the properties of vascular networks in each of these cases impacts on transported molecules, that may be considered as representative "biomarkers", thus making them useful in pathophysiology-assessment.

2 Methods: Key Concepts and Equations

The objective of this section is to briefly introduce the concepts underlying the equations that model flow and transport in this chapter. The section summarizes approaches from different text books (e.g. [6, 12, 16]). Scalar quantities (e.g. the local density) or fields (e.g. the local velocity) in fluid mechanics are usually expressed in Eulerian (or spatial) coordinates in a given coordinate system within the observer's frame of reference. Typically the Eulerian specification for fluids in the body is described in Cartesian coordinates in a fixed reference frame. Conceptually, the continuum descriptions resulting from the conservation of mass, momentum and energy, lead to equations for local densities. They can be viewed as generalisations of point mechanics, which describe how to take into account the spatial extension of a body. Another view to arrive to these equations is to characterize a cloud of particles by observables that emerge from local averaging instead of tracking the individual trajectory of each particle (as e.g. fluid particles).

2.1 Continuum Mechanics of Fluids: The Balance of Mass and Momentum

2.1.1 Mass Balance

Consider a case where there is no production or loss of matter. The equation for conservation of mass in its integral form, states that:

$$\frac{dM}{dt} = 0 \text{ with } M = \int_{V(t)} \varrho(\mathbf{x}, t) d^3\mathbf{x}, \tag{1}$$

whereby $\varrho(\mathbf{x}, t)$ denotes the density of mass points in a sufficiently small volume element $d^3\mathbf{x} = dV$ in the volume $V(t)$.[1] $V(t)$ denotes a material volume moving with the fluid enclosing the same set of particles of total mass M : imagine a set of fluid particles enclosed by a non-constraining bag, impermeable to particles. The time differentiation of the integral with a time-dependent geometric volume $V^*(t)$, is performed according to the Reynolds transport theorem, which states that for a continuous and continuously differentiable tensor field $F(\mathbf{x}, t)$, the integral

$$\frac{d}{dt} \int_{V^*(t)} F(\mathbf{x}, t) dV = \int_{V^*(t)} \frac{\partial}{\partial t} F(\mathbf{x}, t) dV + \int_{A^*(t)} F(\mathbf{x}, t) \mathbf{v} \cdot \mathbf{n} dA, \tag{2}$$

whereby $\mathbf{v}$ is the velocity of the volume element $dV^* = V^*(t + \Delta t) - V^*(t)$, which is the same as the one of its boundary dA^*, and $\mathbf{n}$ the local outwards pointing normal to its boundary $\mathbf{dA} = dA\mathbf{n}$. This equation holds for any chosen volume $V^*(t)$ (which could also be a control volume, i.e. a constant *frozen* volume, fixed at a certain position). The first term on the right-hand side (rhs.) denotes the quantity of F changing with time at the current location, while the 2nd term is the flux passing through the surface $A^*(t)$.

For example, for the mass,[2] $F = \varrho$,

$$\frac{d}{dt} \int_{V^*(t)} \varrho(\mathbf{x}, t) dV = \int_{V^*(t)} \frac{\partial}{\partial t} \varrho(\mathbf{x}, t) dV + \int_{A^*(t)} \varrho(\mathbf{x}, t) \mathbf{v} \cdot \mathbf{n} dA. \tag{3}$$

---

[1] Often instead of d/dt the notation D/Dt is used to specify a time derivative in the material (synonym: Lagrangian) frame.

[2] Often the equation is formulated for the choice $F(\mathbf{x}, t) = \varrho(\mathbf{x}, t) f(\mathbf{x}, t)$ in which case the choice $f = 1$ yields the equation for mass balance, and the last term on the rhs. of Eq. (2) denotes the flux of quantity $\int_{V^*(t)} \varrho(\mathbf{x}, t) f(\mathbf{x}, t) dV$ through $A^*(t)$.

The total change of mass is controlled by a temporal change of mass density inside the volume $V^*(t)$ plus the mass transported through the border $A^*(t)$. If the volume moves with the fluid flow of velocity $\mathbf{u}(\mathbf{x}, t)$ i.e., if the enclosed volume is a material volume, then $V^*(t) = V(t)$, and Eq. (3) becomes by virtue of Eq. (1):

$$\int_{V(t)} \frac{\partial}{\partial t} \varrho(\mathbf{x}, t) dV + \int_{A(t)} \varrho(\mathbf{x}, t) \mathbf{u} \cdot \mathbf{n} dA = 0. \tag{4}$$

Using the divergence (or Gauss) theorem, $\int_V \nabla \cdot \mathbf{Q} dV = \int_A \mathbf{Q} \cdot \mathbf{n} dA$, the area integral in the last equation can be transformed into a volume integral:

$$\int_{V(t)} \left\{ \frac{\partial}{\partial t} \varrho(\mathbf{x}, t) + \nabla \cdot (\varrho(\mathbf{x}, t) \mathbf{u}(\mathbf{x}, t)) \right\} dV = 0. \tag{5}$$

This equation has to be fulfilled for any choice of the integration volume, requiring the integrand to vanish:

$$\frac{\partial}{\partial t} \varrho(\mathbf{x}, t) + \nabla \cdot (\varrho(\mathbf{x}, t) \mathbf{u}(\mathbf{x}, t)) = 0. \tag{6}$$

This is the continuity equation, the local form of mass balance, in the Eulerian description.

2.1.2 Momentum Balance

The (linear) momentum balance in the continuum corresponds to Newton's second law in point mechanics i.e., $d/dt(m\mathbf{v}) = \sum_i \mathbf{f}_i$, expressing that the rate of change of momentum of a particle of mass m and velocity $\mathbf{v}$ is determined by all forces $\mathbf{f}_i$ acting on it. In continuum mechanics this translates for the material volume V into

$$\frac{d}{dt} \int_{V(t)} \varrho(\mathbf{x}, t) \mathbf{u}(\mathbf{x}, t) dV = \int_{V(t)} \varrho(\mathbf{x}, t) \mathbf{b}(\mathbf{x}, t) dV + \int_{A(t)} \mathbf{t} dA, \tag{7}$$

whereby $\mathbf{b}(\mathbf{x}, t)$ denotes the body force per unit of mass, and $\mathbf{t} = d\mathbf{f}/dA$ the surface traction, i.e. a contact or external surface force $\mathbf{f}$ per unit of area in the current deformed configuration. The stress vector can be written as $\mathbf{t} = \mathbf{T} \cdot \mathbf{n}$ with $\mathbf{T}$ the Cauchy stress tensor. In component form for each i and j coordinate and following the Einstein summation convention, it reads $t_i = T_{ij} n_j$. T_{ij} denotes the stress in direction j acting on a surface with normal vector i. $i = j$ denotes normal (compressing or dilatation) stress, and $i \neq j$ shear stress.

The momentum balance Eq. (7) is transformed a. by reformulating the lhs. of Eq. (2) by virtue of the Reynolds transport theorem (Eq. (2)) with the choice

$F(\mathbf{x}, t) = \varrho(\mathbf{x}, t)\mathbf{u}(\mathbf{x}, t)$, b. by following the same steps as for the derivation of the continuity equation (6), and c. by employing the latter equation, into:

$$\varrho(\mathbf{x}, t)\frac{\partial \mathbf{u}(\mathbf{x}, t)}{\partial t} + \varrho(\mathbf{x}, t)\mathbf{u}(\mathbf{x}, t) \cdot \nabla\mathbf{u}(\mathbf{x}, t) = \varrho(\mathbf{x}, t)\mathbf{b}(\mathbf{x}, t) + \nabla \cdot \mathbf{T}, \quad (8)$$

with $[\nabla \cdot \mathbf{T}]_j = \partial T_{ij}/\partial x_i$.

2.1.3 System of Equations of a Continuum

Balance equations for angular momentum and energy can be formulated equivalently. However, since they are not required here, we refer to the relevant literature [12, sections 7.5 – 7.6].

The equations for mass and momentum balance ((6) and (8)) yield 4 equations for unknowns ϱ, $\mathbf{u}$, $\mathbf{T}$. The considered body force is often the gravity force. In order to close the system of equations for mass and momentum balance, an equation is needed that specifies the stress tensor $\mathbf{T}$ according to the property of the considered material. The tensor must fulfill a number of principles (causality, determinism, local action, fading memory, equi-presence, material frame indifference or objectivity, material symmetries, dissipation).

2.2 *Hemodynamics Modeling*

In solids stress is proportional to strain, but in fluids stress is proportional to strain rate. The strain rate tensor is a symmetric tensor given by $[\mathbf{d}]_{ij} = \frac{1}{2}\left(\frac{\partial u_j}{\partial x_i} + \frac{\partial u_i}{\partial x_j}\right)$, expressing the local rate of deformation of the group of particles within a small volume.

2.2.1 The Navier-Stokes Equations

Fluids have in general two stress components: the elastic stress, generated by volume change, and the viscous stress which is the response of the fluid resisting deformation. If the fluid is isotropic, homogeneous, and the effect of temperature negligible, the stress tensor typically reads

$$\mathbf{T} = -p\mathbf{I} + \lambda tr(\mathbf{d})\mathbf{I} + 2\mu\mathbf{d}. \quad (9)$$

Here, $-p$ is defined as the average normal stress, λ a pre-factor and μ the dynamic (or shear) viscosity. Combining Eqs. (9) and (8) leads to the Navier-Stokes equations (for simplicity, the argument $(\mathbf{x}, t)$ is dropped):

$$\varrho\left[\frac{\partial \mathbf{u}}{\partial t} + \mathbf{u} \cdot \nabla\mathbf{u}\right] = -\nabla p + \lambda\nabla(tr(\mathbf{d})) + 2\nabla \cdot (\mu\mathbf{d}) + \varrho\mathbf{b}. \quad (10)$$

This definition of pressure leads to $\lambda = -2\mu/3$ (the trace of the deviatoric stress should be zero).[3]

Blood mainly consists in water and hence can be assumed as incompressible, i.e. $\nabla \cdot \mathbf{u} = 0$. To be consistent with a homogeneous fluid, ϱ is assumed homogeneous. Thus blood obeys the Navier-Stokes equations for an incompressible fluid:

$$\varrho \left[\frac{\partial \mathbf{u}}{\partial t} + \mathbf{u} \cdot \nabla \mathbf{u} \right] = -\nabla p + \nabla \cdot \left(\mu (\nabla \mathbf{u} + \nabla \mathbf{u}^{\mathbf{T}}) \right) + \varrho \mathbf{b}. \tag{11}$$

Note that μ at this point can either be a constant (or a function independent of the fluid variables), in which case blood is considered a Newtonian fluid, or can be a function of the shear rate or other fluid variables, defining it as non-Newtonian.

2.2.2 The (Hagen-)Poiseuille Flow Model

Microcirculatory networks such as the small artery trees feeding organs or the sinusoidal network in the liver, can be viewed as a system of one dimensional circular pipes with any orientations and branchings, so that the gravity force can take in each pipe a different direction. Combined with its relatively small contribution at this mesoscale, the gravity force is often neglected. Moreover, these vessels are of small diameter (e.g., arterioles and sinusoids have a diameter of about 300–15 μm and 15–4 μm respectively), quite stiff and connected to root vessels where flow pulsatility is damped. Thus, these pipes are modeled here as rigid, impermeable to fluid leakage, with a unidirectional, axisymmetric and stationary flow ($\partial \mathbf{u}/\partial t = \mathbf{0}$).

Consider a pipe element of length L and radius R, oriented towards the z-direction of its centerline, for which cylinder coordinates are a natural choice. The assumptions above lead to $u_\theta = u_r = 0$, and since $\nabla \cdot \mathbf{u} = 0$, eventually $\partial u_z / \partial z = 0$, and $u_z = u_z(r)$ only. The Navier-Stokes equations thus simplify to:

$$\begin{cases} -\dfrac{\partial p}{\partial z} + \dfrac{1}{r}\dfrac{\partial}{\partial r}\left(\mu r \dfrac{\partial u_z}{\partial r} \right) = 0 \\ \dfrac{\partial p}{\partial \theta} = \dfrac{\partial p}{\partial r} = 0 \implies p = p(z) \end{cases}$$

Hence, by integrating twice over r with the hypotheses above leading to $u_z(r = R) = 0$ and considering a finite velocity at the center $r = 0$, $u_z = -\frac{1}{4\mu}\frac{dp}{dz}(R^2 - r^2)$, R being the radius of the pipe. This is the equation of a parabolic profile, with the

[3] An alternative definition of the pressure to be the thermodynamic equilibrium pressure p_e, leads to the same equations but written in terms of p_e, κ the bulk viscosity and μ, with $p = p_e - \kappa \nabla \cdot \mathbf{u}$.

maximum velocity at its center, oriented along the direction of the vessel. Integration of the flow speed over the cross section of the pipe yields the flow rate:

$$Q = \int_0^{2\pi} \int_0^R u_z r dr d\theta = -\frac{\pi R^4}{8\mu} \frac{dp}{dz} = -\frac{\pi R^4}{8\mu} \frac{P_{out} - P_{in}}{L}, \tag{12}$$

P_{in} being the pressure at the inlet of the pipe, P_{out} at the outlet of the pipe. In short form, this equation can be written as the well-known *Poiseuille law*:

$$P_{in} - P_{out} = \mathcal{R}Q, \tag{13}$$

where $\mathcal{R} = (8\mu L)/(\pi R^4)$ denotes the resistance to flow. The obtained Poiseuille law is an analogy to the law relating electrical current and electrical tension via the electrical resistance. This analogy is made to construct hemodynamics models that only depend on time, called *0D* models, and can further be extended to the entire circulation to include elasticity of the vessel walls, fluid inertial effects, valves that open and close like diodes, etc. (e.g. [1]).

The wall shear stress (i.e. the tangential part of the stress vector at the wall of normal $\mathbf{n}$) is then:

$$\|\mu(\nabla\mathbf{u} + \nabla\mathbf{u}^\mathbf{T}) \cdot \mathbf{n} - \mathbf{n} \cdot \mu(\nabla\mathbf{u} + \nabla\mathbf{u}^\mathbf{T}) \cdot \mathbf{n}\| = \frac{4\mu Q}{R^3}. \tag{14}$$

Because endothelial cells lining the vessel wall are sensitive to wall shear stress, this quantity is computed for modeling a number of mechano-biology processes, such as angiogenesis, vessel remodeling or thrombosis development.

2.3 Viscosity Law

Note that in the derivation above, the viscosity μ is independent of r, θ but can depend on z. The simplest way of describing blood flow properties is to consider a *Newtonian fluid* with a constant viscosity μ, which is the same throughout the network. While this model is sufficient in some applications, the importance of heterogeneous nature of blood at the considered scale calls for a more complex description. One way of accounting for the effects of red blood cells on the flow properties in each pipe segment is to prescribe an *apparent viscosity*. One effect, the *Fåhræus-Linqvist effect*, is the lowering of the viscosity for a decreasing radius (from 300 μm down to 10 μm), as red blood cells are transported in the middle of the vessels where velocity is higher, and then an increased viscosity for smaller vessels.

Pries and Secomb [17] proposed an empirical formula that changes the apparent viscosity according to the radius of the vessel R:

$$\mu(R)=\mu_{\mathrm{P}}\left[1+(\mu_{45}(R)-1)\frac{(1-H_{\mathrm{D}})^{C(R)}-1}{(1-0.45)^{C(R)}-1}\left(\frac{2R}{2R-1.1}\right)^2\right]\left(\frac{2R}{2R-1.1}\right)^2, \tag{15}$$

where μ_{P} is the blood plasma viscosity and H_{D} is the hematocrit discharge. $C(R)$ is a scaling exponent that is about -0.8 for vessel radius above 20 μm and constant for smaller vessels. $\mu_{45}(R)$ is the apparent viscosity at $H_{\mathrm{D}} = 0.45$. Its empirical formula is $\mu_{45}(R) = 6e^{-0.17R} + 3.2 - 2.44e^{-0.06(2R)^{0.645}}$ (with R in μm). Other effects can be taken into account such as red blood cells partition that deviates from the flow rate partition at bifurcations but will not be considered in the following (see [2]).

In the next sections transport equations are introduced, since blood or other biofluids such as bile, can transport different injected species.

2.4 *Transport in a Vascularized Tissue*

2.4.1 Multi-Components, Single Phase Models

Transport of one type of particles in something else than void requires consideration of at least two components in a so called *mixture*.

A well-known example is that of Brownian particles at low density in a fluid. Each Brownian particle, if released in a flow performs a movement composed of advection and diffusion i.e., a directed movement with the average flow speed of the fluid overlaid with random displacements. The driving force for this is collisions with the fluid particles (typically water). Examples addressed below are a metabolite and a contrast agent, molecules transported with the blood but, different from the Brownian particle, not orders of magnitude bigger than the solvent molecules. Hence, the necessity to consider *multicomponent (or multispecies) models*. The total density ϱ of a multicomponent fluid of volume V is given by the sum over all components $i = 1, .., n$ of the partial densities $\varrho_i = m_i/V$, defined from their respective mass m_i: $\varrho = \sum_i^n \varrho_i$. From them, a mass fraction can be defined as $\phi_i = \varrho_i/\varrho$, fulfilling $\sum_i^n \phi_i = 1$. The mass flux of component i then becomes $\mathbf{J}_i = \varrho_i \mathbf{u}_i$, $\mathbf{u}_i$ being the velocity of each component.

As molecules could in principle undergo chemical reactions in the blood, they could in principle be generated or annihilated (leading to a "birth-death" process), hence the mass of a particular species is not necessarily conserved. Mass balance for each individual component therefore becomes (dropping the arguments "$(\mathbf{x}, t)$"):

$$\frac{\partial}{\partial t}\varrho_i + \nabla \cdot (\varrho_i \mathbf{u}_i) = \sum_{j=1}^{r} \nu^{ij} \mathcal{J}_i, \tag{16}$$

whereby $i = 1, \ldots, n$ denotes the species, j the reaction counter. Mass conservation of the entire set of species imposes that $\sum_{i=1}^{n} \sum_{j=1}^{r} \nu^{ij} \mathcal{J}_i = 0$. Consequently, summing in Eq. (16) over all species $i = 1, \ldots, n$ leads again to the mass conservation of the total number of species $\sum_i \varrho_i$ moving with average speed $\mathbf{u}^{\varrho} = (1/\varrho) \sum_i \varrho_i \mathbf{u}_i = \sum_i \phi_i \mathbf{u}_i$. From the mass balance equation for each component Eq. (16), the equation of concentrations can be obtained by $c_i = \varrho_i / W_i$ whereby W_i denotes the specific mass of the considered molecule, leading after some rearrangements to

$$\frac{\partial c_i}{\partial t} + \nabla \cdot \big(c_i (\mathbf{u}_i - \mathbf{u}^{\mathbf{c}})\big) + \nabla \cdot (c_i \mathbf{u}^c) = \sum_{j=1}^{r} \nu^{ij} \mathcal{J}_i / W_i, \tag{17}$$

whereby the total concentration c in a mixture volume V is defined as $c = \sum_i N_i / V = \sum_i c_i$, and the number of particles N_i of a component relates to its mass m_i as $N_i = m_i / W_i$, respectively. The 2nd term $J_i^c = (c_i (\mathbf{u}_i - \mathbf{u}^{\mathbf{c}}))$ on the lhs. denotes the molar diffusion flux, representing the difference in advection of component i due to its specific flow velocity $\mathbf{u}_i$ compared to the advection one would observe if the component would be transported with the concentration-based average flow velocity $\mathbf{u}^c = (1/c) \sum_i c_i \mathbf{u}_i = \sum_i \omega_i^c \mathbf{u}_i$.

To determine how the flow velocity of a certain component deviates from that of the average, which is usually controlled by the flow speed of the solvent (the blood), possible causes for such a flux need to be determined. This emerges from the relation of the entropy production rate, in which fluxes and their conjugate forces appear as products in each term. The emerging reciprocity relations between generalized forces and generalized fluxes can be of scalar, vectorial or tensorial type, whereby each force and the flux it generates need to be of the same tensorial order. Generalized forces contributing to particle flux are a gradient of the chemical plus non-chemical potential, and the gradient of temperature. In case of a diluted ideal mixture, and in absence of a temperature gradient, the diffusive (particle) flux is given by Fick's law for diffusion, $\mathbf{J}_i^c = -D_i \nabla c_i$, D_i being the coefficient of diffusion. Hence, the balance equation for the concentration of species i becomes:

$$\frac{\partial c_i}{\partial t} + \nabla \cdot (c_i \mathbf{u}^c) - \nabla \cdot (D_i \nabla c_i) = \sum_{j=1}^{r} \nu^{ij} \mathcal{J}_i / W_i. \tag{18}$$

This transport equation mimics the transport of a molecular species i with the solvent. Often however, a considered volume element contains not only the fluid phase but also other phases, for example, a blood vessel with its enclosing interstitium. This requires to separate in each micro-region between different spatial phases in addition to the separation of species in a mixture. This is the subject of the next section.

2.4.2 Multi-Components, Multi-Phases Models

Often, one zooms out at larger scales (mesoscale) such that a representative volume element (RVE) contains not only one phase (e.g. blood) but also other phases (e.g. interstitium, tissue cells). In this case the equations at mesoscale differ from that at microscale in that they contain inner phase boundaries that need to be taken into account. Consider first for simplicity the case where no transition of species i between different phases occurs, e.g. a contrast agent transported with the blood whereby the vessel borders are impermeable for the contrast agent, also called an *intravascular* contrast agent. For each individual phase p i.e., at microscale, equations (Eqs. (17) and (18)) now read in their respective 3D domain Ω^p,

$$\frac{\partial \varrho_i^p}{\partial t} + \nabla \cdot (\varrho_i^p \mathbf{u}^{c,p}) - \nabla \cdot \left(D_i^p \nabla \varrho_i^p\right) = \sum_{j=1}^{r} \nu^{ij} \mathcal{J}_i^p, \tag{19}$$

and

$$\frac{\partial c_i^p}{\partial t} + \nabla \cdot (c_i^p \mathbf{u}^{c,p}) - \nabla \cdot \left(D_i^p \nabla c_i^p\right) = \sum_{j=1}^{r} \nu^{ij} \mathcal{J}_i^p / W_i. \tag{20}$$

The volume dV of the RVE in the entire 3D space $\Omega = \cup\Omega^p$ is chosen: this defines the mesoscale. Equation $\phi^p = dV^p/dV$ defines the volume fraction of phase p in the RVE. Observable densities refer to their microscale-averages in the volume of the RVE at mesoscale. One thus operates with the quantity $\phi^p \bar{\varrho_i}^p = \phi^p (dm_i^p/dV^p) = dm_i^p/dV$. In the following, to not make notations heavy, we drop the bar for all unknowns and parameters. This leads by spatial integration over this small RVE volume when the geometry of each phase is constant in time, to the following equations at mesoscale [16]:

$$\frac{\partial(\phi^p \varrho_i^p)}{\partial t} + \nabla \cdot (\phi^p \varrho_i^p \mathbf{u}^{c,p}) - \nabla \cdot \left(D_i^p \phi^p \nabla \varrho_i^p\right) = \sum_{j=1}^{r} \phi^p \nu^{ij} \mathcal{J}_i^p, \tag{21}$$

and

$$\frac{\partial(\phi^p c_i^p)}{\partial t} + \nabla \cdot (\phi^p c_i^p \mathbf{u}^{c,p}) - \nabla \cdot \left(D_i^p \phi^p \nabla c_i^p\right) = \sum_{j=1}^{r} \phi^p \nu^{ij} \mathcal{J}_i^p / W_i. \tag{22}$$

Note that at this stage, the precise contours of the interfaces between different phases are not resolved anymore. Hence the particle flux previously defined, $\mathbf{J}_i^c$ represents both a diffusion process and a dispersive effect. The former is actually an effective diffusion, since diffusion might be hindered by the complex microscale geometry that is no longer resolved (e.g. diffusion in the interstitial space between the capillaries and the cells). Dispersion is here a mixing due to the non-homogeneous microscale mixture velocity. Both phenomena can actually take the same mathematical form of a diffusion flux [16], so that $\mathbf{J}_i^{c,p} = -D_i^p \nabla c_i^p$. This needs to be verified based on the given sub-tissue microscale geometry.

The derivation of the mesoscale equations in case of permeable phase interfaces is lengthy [16], leading to a similar equation as for the microphase:

$$\frac{\partial(\phi^p \varrho_i^p)}{\partial t} + \nabla \cdot (\phi^p \varrho_i^p \mathbf{u}^{c,p}) - \nabla \cdot \left(D_i^p \phi^p \nabla \varrho_i^p\right) = \sum_{j=1}^{r} \phi^p \nu^{ij} \mathcal{J}_i^p + \sum_{p' \neq p} e_{pp'}^{ip}, \quad (23)$$

whereby $e_{pp'}^{ip}$ denotes the transfer of species i between phases p and p', with $e_{pp'}^{ip} > 0$ if the transfer is from the phase $p' \to p$. For the concentration $c_i^p = \varrho_i^p / W_i$,

$$\frac{\partial(\phi^p c_i^p)}{\partial t} + \nabla \cdot (\phi^p c_i^p \mathbf{u}^{c,p}) - \nabla \cdot \left(D_i^p \phi^p \nabla c_i^p\right) = \sum_{j=1}^{r} \phi^p \nu^{ij} \mathcal{J}_i^p / W_i + \sum_{p' \neq p} e_{pp'}^{ip} / W_i. \quad (24)$$

The considered transfers for an injected substance in the body between tissue components are functions of the concentration in the phase from which the specie is transferred when the transfer is performed *actively* by molecular transporters, or of concentration differences across phases when a specie crosses the blood vessel endothelial layer or cell membrane *passively* due to diffusion across the *membrane*, e.g. $e_{pp'}^{ip} / W_i = k_{pp'}^{ip}(c_i^{p'} - c_i^p)$. Active transfers are typically linear, or nonlinear, modelled by a Michaelis-Menten term to capture saturation effects $e_{pp'}^{ip} / W_i = a_i^p c_i^p / (b_i^p + c_i^p)$, where a_i^p and b_i^p are constants [2]. In both cases, these transfers are proportional to the interphase interface area. These transfers will be specified for each case in the Results section.

Finally, the mesoscale equations revert to the microscale ones if only one phase p is present such that $\phi_p = 1$, and the RVE scale is the same as the microscale.

2.4.3 Compartment Models

Let us consider a certain ROI, which can be of the order of a RVE (at a few cells or capillaries) or larger than this (a tissue piece that corresponds to a DCE-PI imaging voxel or multiple voxels). Within this ROI, assuming that the concentration is spatially homogeneous within an ROI, i.e. making the **well-mixed compartment** hypothesis [18], Eq. (24) can be simplified as an ordinary differential equation on the average concentration in a phase c_i^p connected to another phase $p' \neq p$:

$$\phi^p \frac{\mathrm{d}c_i^p}{\mathrm{d}t} = \sum_{j=1}^{r} \phi^p \nu^{ij} \mathcal{J}_i^p / W_i + \sum_{p' \neq p} \left(e_{pp'}^{ip} / W_i \right). \tag{25}$$

This leads then to a **multi-compartment** model, each phase p being a different compartment.

When discretizing in space the tissue around the vessel networks, the discrete PDE multi-phase model can be seen as neighboring compartments (of the same phase) connected to each other: these different ROIs are connected to each other. A complete derivation of this equation is performed in appendix. If for example the transport is driven by flow with velocity $\mathbf{u^{c,p}}$ such that $\mathbf{u^{c,p}A^p} = Q^{c,p}$ denotes the flow rate exiting a compartment, $\Sigma_{p' \neq p} e_{pp'}^{ip} / W_i = \frac{1}{V}(\sum_{p', Q^{c,p'p}>0} Q^{c,p'p} c_i^{p'} - Q^{c,p} c_i^p)$, i.e., the transfer of species i from phase p to neighbor phase p' is proportional to the fluxes.

2.5 *Vascular Networks*

The domain of study is a vascularized tissue, for which each phase and component need to be defined. In order to understand the influence of specific geometries on the flow and transport of different injected substances, a microvascular network is constructed. This is a choice made to build the multiphase model at the mesoscale, with known interface conditions across the different phases.

2.5.1 Geometry

The studied micro-vasculature is modelled by a network of cylinders of different radii, which are edges of a directed graph connected to each other by nodes. Microcirculatory networks of vessels are difficult to image and segment in an extent such that one knows which boundary conditions to apply. Another route is to generate such networks based on algorithms that combine morphometric rules only, or combined with fluid mechanics principles. Morphometric rules typically respect statistics of real segmented microvasculatory networks.

The baseline networks for a normal tissue in Case 1 (Sect. 3.1), are constructed based on the following algorithm: a first network is generated (Algorithm 1) and then remodeled to be realistic (Algorithm 2). The idea here is to construct arterial and venous trees connected by capillaries. The baseline network rules are then changed to model a vascularized tumor network (Case 1, Sect. 3.1). For a more complete description, see [21].

The liver lobule vascular network of cases 2 and 3 is generated based on statistics of real liver lobules (see [2, 5]).

Algorithm 1 Generation of the network

Input: (x, y), Size of the network
Output: Realistic vascularization

1. Place a root artery e.g. in the middle of the domain
2. Place root veins e.g. at the borders of the network until there are enough
3. Start a random growth that fills the entire domain
4. Add capillaries between the remaining adjacent tips
5. **while** step < N_{steps}, **do**

    Shear-stress homogenization
    Increase step by 1

   end while

Algorithm 2 Shear-stress homogenization

Input: Vessel network
Output: Homogenized vessel network

1. Remove capillaries to enable new sprouts
2. Sprout new vessels from each node with free neighbours, with probability p_{spr}
3. Add capillaries between adjacent tips. Set the capillary radius to `MIN_RADIUS` and the microvascular density n for the region
4. Update the radii starting from the capillaries' neighbours.

$$r_{i,j}^{\alpha} = n \sum_{k,k \neq j} r_{i,k}^{\alpha}$$

5. Compute the pressure field in the entire network with Eq. (26)
6. Compute the wall shear-stress τ in each vessel with Eq. (14)
7. Collapse all vessels with probability $p_{coll} = \frac{\tau_{max} - \tau}{\tau_{max} - \tau_{min}}$

In this chapter, we are interested in the effect of varying geometry on the resulting hemodynamics and transport of a species diluted in blood, with possible influences on metabolism. Therefore, variations of the baseline network of the test cases will be defined for each case in Sect. 3.

2.5.2 Hemodynamics in Vessel Networks

The Poiseuille flow model is derived above for a single vessel. A vessel network is a connected, directed, graph of vessels, where blood flow rate Q and pressure P are solved for as follows.

Hemodynamics Variables in a Network of Cylinders

Let us consider the vessel (i, j) connecting nodes i and j

- P_i the pressure at node i
- $Q_{i,j}$ the flow through vessel (i, j)
- $\mathcal{R}_{i,j} = \frac{8\mu_{i,j} l_{i,j}}{\pi R_{i,j}^4}$ the Poiseuille resistance to flow of vessel (i, j). Note that blood viscosity μ depends on the radius of the vessel (i, j) (Eq. 15).

In each vessel (i, j), Eq. (13) relates the flow through it to the pressure loss between its entry node j and its ending node i. In addition mass is conserved at each bifurcation or multifurcation (the flow rate entering a bifurcation is equal to the flow rate exiting a bifurcation), and pressure is modeled as continuous (there are no pressure losses due to 3D effects at that scale, although it can be included if needed for the larger vessels).

$$\left.\begin{aligned} P_i - P_j = \mathcal{R}_{i,j} Q_{i,j} \\ \forall i \in \mathcal{N}, \sum_j Q_{i,j} = 0 \end{aligned}\right\} \Rightarrow P_i \sum_j G_{i,j} = \sum_j P_j G_{i,j} \tag{26}$$

$$G_{i,j} = \frac{1}{\mathcal{R}_{i,j}}$$

This leads to a linear system of equations on the pressures $(P_i)_{i\in\mathcal{N}}$ relating the pressure at each node i to the pressures at the neighbouring points j. The system is solved due to the specification of *boundary conditions*, typically pressure, flow rate or a function of them, one condition per graph root or end node, one of them being not a flow rate condition in order to respect mass conservation. They depend on the specific study case (see Sect. 3).

2.5.3 Transport

Species are injected in the system, then transported throughout the vessel network. They possibly leak out of vessel walls in the smaller vessels of the network and finally leave the system by going back into the network and exiting it, or by being metabolised by cells. Depending on the study case, some terms in Eq. (24) are indeed neglected for each phase of the system. Studying how the concentration time profile changes from the input to the network's output, yields valuable insights into the network itself. Finally, the resulting equations (purely advection, advection-reaction, etc) are then discretized on the network by different numerical techniques depending on the study case.

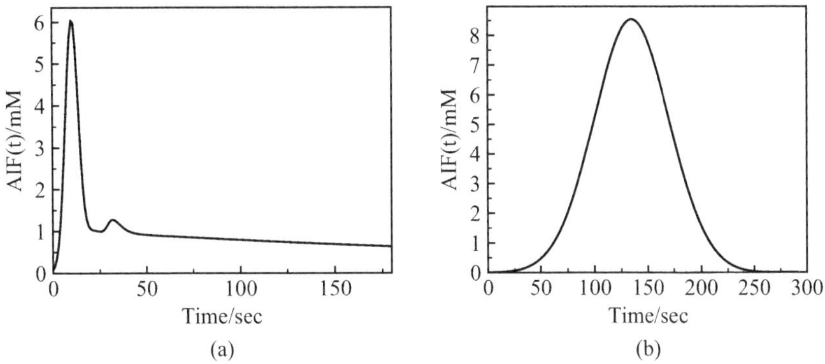

Fig. 1 Shape of the two arterial input functions, inlet boundary condition for transport in vascular networks. (**a**) From Parker. (**b**) Gaussian

Inlet Boundary Conditions

Transport in the vascular phase is typically an advection or advection-reaction equation. Hence, one boundary condition in the vascular network is required at each inlet. In this chapter, we study the effect of different input functions (or arterial input functions—AIF). The first one (Fig. 1a) comes from [15] and represents an average concentration from DCE-PI patients, entering an organ arterial vasculature. The second one (Fig. 1b) is a Gaussian function that can represent an arterial input function in the case of a very long injection time, such as intra-peritoneum injection for instance [5]. Double exponential functions are also emerging from metabolic processes and are considered for local AIF to a functional unit (see Results in Sects. 3.2 and 3.3).

Characterization of a Network: Mean Transit Time

As underlined in the introduction, a simple biomarker that permits to assess perfusion of a region of interest (ROI) is the *mean transit time*, defined as the average time it takes for an input unit particle of matter to travel across the ROI. See [14] or [2] for more background.

For a network $\mathcal{N}$ in a given ROI with $\mathcal{I} \subset \mathcal{N}$ the set of input nodes and $\mathcal{O} \subset \mathcal{N}$ the set of output nodes of the network, if Q_i is the blood flow rate at the input $i \in \mathcal{I}$, Q_o at the output $o \in \mathcal{O}$ and q is the total injected quantity ($q = \sum_{i \in \mathcal{I}} \int Q_i c_i \mathrm{d}t$), the mean transit time is computed by

$$\overline{t_T} = \sum_{o \in \mathcal{O}} \frac{\int_{-\infty}^{+\infty} t\, Q_o c_o \mathrm{d}t}{q} - \sum_{i \in \mathcal{I}} \frac{\int_{-\infty}^{+\infty} t\, Q_i c_i \mathrm{d}t}{q} \tag{27}$$

From the definition of the total injected quantity q and assuming conservation of the species' mass across the network $\left(\sum_{i \in \mathcal{I}} \int Q_i c_i \mathrm{d}t = \sum_{o \in \mathcal{O}} \int Q_o c_o \mathrm{d}t\right)$, the mean transit time represents the expected value of the time spent inside the network: the expected value of the time it enters is subtracted from the expected value of the time it exits.

Numerical Schemes and Algorithms
To integrate the PDEs and ODEs over time on the studied vascular network, one must approximate the exact solution by a numerical solution in each vessel and connect them at bifurcations. Elliptic equations, such as steady state oxygen supply, are receiving a renewed attention for very large, multiscale networks [20]. Solving hyperbolic equations on a network to capture time-varying phenomenon, is yet another challenge, and more challenging than on single vessels [2]. Most of the discretization steps considered in this chapter are presented in the following publications: [2, 5, 21].

Moreover, as depicted in Eq. (27), the mean transit time evaluation requires an integration of a certain quantity to infinite times. The integration method used here is the trapezoidal rule. However in practice (and it is true for *in-silico* as well as *in-vivo* imaging signal), the acquisition time is finite and much smaller than the clearance time in certain tissues. The mean transit time might then be under-estimated. To get rid of this bias, the output signal is extrapolated by a decaying exponential following the final trend of the simulated signal.

3 Results

In the ensuing section, we elucidate the application of the aforementioned models within three prototypical case studies concerning hemodynamic circulation and mass transport in extensive vascular networks. The first case explores dynamic imaging analysis for tumor assessment, the second one pertains to metabolism in a liver lobule, while the third one presents modeling of drug toxicity. For each study, different geometries of microcirculation are built in order to better understand the effect of the network geometry on blood flow and transport of a diluted species.

3.1 Case 1: Vascularized Tumor

DCE-PI, such as DCE-MRI, represents an imaging modality employed primarily for investigating perfusion evaluation within specific regions of significance in an organ. Indeed, the concentration of a contrast agent (here supposedly proportional to the received signal intensity without loss of generality) is acquired over time in a 2.5D slice of tissue (Fig. 2). The signal can exhibit abnormal behaviors such as in the presence of a tumor: in Fig. 2b, the tumor is highlighted in the bottom left corner.

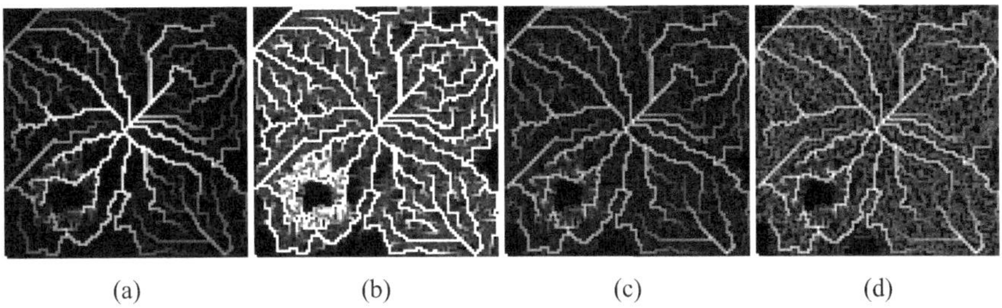

Fig. 2 Example of *in-silico* contrast-enhanced MRI images over four time points. The contrast agent enters through small arteries (**a**), goes into capillaries (**b**) and diffuses out of them (**c, d**), before leaving the tissue through small veins. The tumor *lights up* here due to its enhanced micro-vasculature

3.1.1 Motivations

By modeling the behavior of the contrast agent over time in a certain region of interest, it is possible to extract its perfusion and permeability features in order to assess the characteristics of this region [18]. Determining the parameters of a model from provided series of images is referred to as the inverse problem.

We here present an *in-silico* model of a tissue micro-vasculature that helps in understanding the effect of a change in geometry—such as the development of a tumor—on the perfusion of this tissue. The model has been presented in details and applied to a wide range of 1D, 2D and 3D cases in [21].

3.1.2 Specific Equations and Boundary Conditions

In this specific example, the injected species is diluted in the blood plasma and can leak outside of the vessels in the interstitial space. Equation (24) is applied on both the plasma (indexed P) and interstitium phases (indexed I) by considering the following processes:

- In the blood, the agent is transported by blood flow and can leak through the walls of the capillaries because of their permeability.
- In the interstitial space, the agent is transported thanks to diffusion and can return to blood through the capillary walls.
- The exchange across the vessel walls is supposed passive here, so driven by the concentration difference across the walls.

In practice these assumptions are taken into account by setting $p \in \{P, I\}$, $\nu^{i,j} = 0$, $D^P = 0$, $u^I = 0$, $e^{ip}_{pp'}/W_i = k^{PI}(c^p - c^{p'})$ and letting go of the subscript i in Eq. (24). These changes lead to an advection-reaction equation for the plasma concentration and reaction-diffusion in the interstitial space. Furthermore the agent enters the domain Ω through the main artery $\partial\Omega_{in}$ according to the arterial input function, and zero flux across the domain borders ($\partial\Omega$) is assumed which leads to the following problem:

$$\begin{aligned}
\frac{\partial(\phi^P c^P)}{\partial t} &= -\nabla \cdot (\phi^P c^P \mathbf{u}) - k^{PI}(c^P - c^I), \\
c^P|_{\partial\Omega_{in}} &= c_{AIF}(t) \\
\frac{\partial(\phi^I c^I)}{\partial t} &= \nabla \cdot (\phi_I D^I \nabla c^I) + k^{PI}(c^P - c^I), \\
\partial_{\mathbf{n}} c^I|_{\partial\Omega} &= 0
\end{aligned} \tag{28}$$

The transfer rate, the permeability k^{PI}, is the product of an intrinsic permeability and the surface area of the underlying vessel walls within an RVE. Here, since the microvasculature is generated, these parameters can be locally computed [21]. The mesoscopic blood flow fluxes also naturally derive in the RVE from this generated vasculature. In this study an RVE is around $(60\,\mu\text{m})^3$.

3.1.3 Numerical Methods

The vascularization is created stochastically with the generation algorithm in Sect. 1. This enables to have different geometry patterns depending on the state of the tissue: tumor tissues have a higher micro-vascular density (represented by n in the shear-stress homogenization algorithm—Sect. 2) whereas necrotic tissues have a higher collapse probability, compared to normal tissue.

Once blood flow is computed in each vessel by Eq. (26), the equations of transport (Eq. (28)) are discretized via a finite volume implicit scheme (see [21]). The obtained concentration maps are then analyzed as being an *in-silico* representation of a "real" DCE-PI acquisition.

The full code for the creation of the synthetic vascularization and the computation of hemodynamics and transport is available on Gitlab: https://gitlab.inria.fr/simbiotx/dyna-imaging-mod.

3.1.4 Results

The generated networks are displayed in Fig. 3. We can see that the presence of the necrotic core leads to the removal of small vessels inside of it and a remodelling of the blood vessels in the vicinity of the tumor. For instance, due to the increased micro-vascular density in the tumor, the main arterioles feeding the tumor (bottom left of Fig. 3b) have a larger vessel radius than they would have without a tumor (bottom left of Fig. 3a).

For the two networks, blood flow and plasma volume fraction maps as well as the concentration map at different time steps are displayed in Fig. 4. First, it

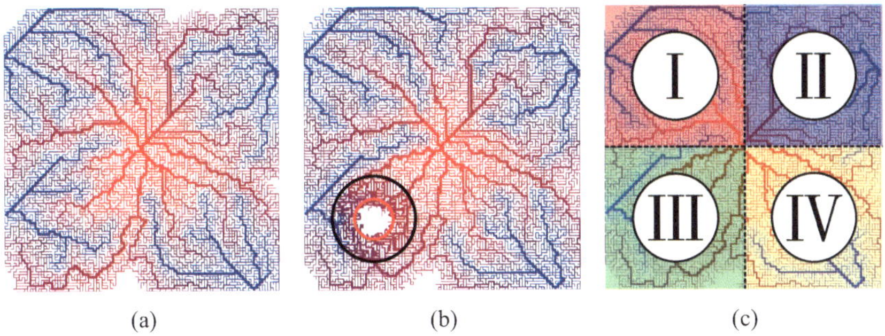

Fig. 3 Generated vascularization and computed pressure field with an imposed pressure gradient between the arterial input and the venous outputs for two different states of the tumor. High pressure vessels are shaded in red while low pressure vessels are represented in blue. (**a**) Without a tumor. (**b**) After adding a tumor and remodelling. The black circle delimits the tumor area. The red circle delimits the necrotic core of the tumor. (**c**) Definition of the four regions of interest (ROI) corresponding to four draining veins

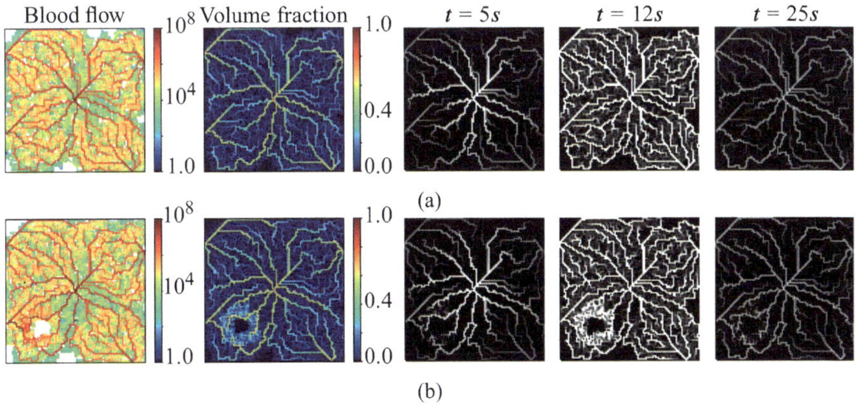

Fig. 4 First column: blood flow map (Q, in $\mu m^3.s^{-1}$). Second column: volume fraction of plasma (ϕ_P, no unit). Third to fifth columns: simulated concentration map in the generated vasculature at different moments in time with Parker's AIF. (**a**) in the case of a healthy tissue. (**b**) with a tumor in the bottom left region, which exhibits both increase vascular density and subsequent higher permeability

is comforting to see that blood flow is higher in voxels (RVEs) of higher plasma volume fraction. We notice that the plasma volume fraction in the low vascularized parts of the tumor region is higher than the lowest vascularized regions of the healthy tissue.

The two input functions (see Fig. 1) are injected in the center of the two generated vascular networks: healthy and tumoral. If you take a look at the different time steps in Fig. 4, at 12 s post-injection, the bottom-left region in the second row, is more densely illuminated than in the first row. There is also a characteristic dark core. This region is indeed that of the tumor: the model reproduces the known behavior

Table 1 Comparison of the transport times for different cases (in seconds). First column: mean transit time of the contrast agent in the tissue. Second to fifth columns: mean transit time of the contrast agent in the four regions of interest (ROI III is the region with the tumor)

Case	AIF	MTT	ROI I	ROI II	ROI III	ROI IV
No tumor	Parker	4.12	3.97	4.59	4.08	3.93
	Gaussian	2.02	1.91	2.26	2.05	1.91
Tumor	Parker	4.05	4.17	4.59	3.44	4.51
	Gaussian	1.93	1.93	2.14	1.72	2.12

of tumor illumination—with tumor necrosis in the center and highly vascularized ring—during contrast agent injection [8, 10]. However, this illumination is quite short in this simulation (around 2 s). Other parameters such as higher intrinsic permeability in the tumor region, generate a different type of illumination [21]: the dynamics of lightening up hence relates to the underlying tissue properties.

Another simple way of detecting a tumor is by looking at the mean transit time of certain regions of interest [11]. The mean transit time of the whole network as well as the one of the four regions of interest defined in Fig. 3c is computed according to Eq. (27). In Table 1, the first two lines show small variations of mean transit times across the four studied areas. In the case of the presence of a tumor, the overall mean transit time is smaller than in the absence of tumor. Furthermore, the mean transit time in the region of the tumor is drastically reduced compared to the baseline whereas the other times are slightly augmented. Indeed, as the tumor is highly vascularized, the blood flow increases. In the tumor region, the high permeability of the blood vessels boosts the exchanges between the interstitium and the plasma and therefore improves the clearance of the tissue, leading to a smaller transit time.

3.2 *Case 2: Lobule*

As presented in the introduction (Sect. 1), an in-depth understanding of liver lobule perfusion characteristics in both healthy and diseased organisms is crucial for pathology assessment and optimal treatment design. This calls for models of microcirculation and transport in liver lobules for different distributions of vessel radii, since their variations are often connected to diseases (cf. [4, 7, 19]). The models in questions should be efficient and practical in the sense that they should capture enough detail in order to characterize the system, as well as being simple enough to allow for efficient numerical implementation. In this section, we present illustrative results of such models on a statistically representative liver lobule with 3 blood flow inlet axes (portal venules, PV) at the vertices of roughly a hexagon of liver capillaries called the sinusoids, and one draining central vein (CV) axis in the center. These models can then be developed further, for example by coupling them to models of exchange with surrounding hepatic cells (hepatocytes) and other organs, as presented in Sect. 3.3.

3.2.1 Specific Equations and Boundary Conditions

The cases presented here are adapted from [2] where the equations as well as numerical implementation are fully described. First, we compute the hemodynamics from Eqs. (1), (7), and (15) with the solution process outlined in Sect. 2.5.2. The same total inlet flow rate was imposed for all simulations, and zero pressure at the PV without loss of generality. Other boundary conditions can be studied [2].

For the transport part, a single chemical compound is considered through the sinusoidal network by pure advection and without exchanges with surrounding hepatocytes, which is equivalent to setting $D = 0$, $\nu^{ij} = 0$, $\phi^p = 1$ and omitting the indices p in Eq. (24), resulting in

$$\frac{\partial c^p}{\partial t} + \nabla \cdot (c^p \mathbf{u}) = 0. \tag{29}$$

This equation can be solved with the finite volume method, by discretizing the vascular network as sub-volumes that partition exactly the vascular network, and integrating the equation above on each of those volumes. When doing so, it is interesting to note that the resulting equation matches the one obtained in appendix (leading to Eq. (25)). Indeed, Eq. (29) can be written in discrete form as

$$V^p \frac{\mathrm{d}\overline{c^p}}{\mathrm{d}t} = \sum_{p' \neq p, Q^{p'p} > 0} Q^{p'p} \overline{c^{p'}} - Q^p \bar{c^p},$$

where $\overline{c^p}$ is the average concentration over a discrete volume V^p and $Q^{p'p}$ are the "inter-compartment" mass fluxes which here correspond to the mass fluxes between volume V^p and the other volumes to which it is connected.

Regarding the AIF, similarly to Sect. 3.3 below, we ran the simulations for a double-exponential profile, however on a shorter time scale for illustrative purposes. The profile is shown for reference in Fig. 7.

3.2.2 Results

The geometries considered are displayed and described in Fig. 5. We will refer to each case as "reference" or "uniform", "Gaussian" and "dilated" or "dilated pericentral" for the remaining results Sects. 3.2 and 3.3. The exact parameters of the distributions are

- $r_{\text{ref}} = 4.66$ μm for the reference distribution,
- a Gaussian distribution with a mean of 4.85 μm and a standard deviation of 1.36 μm,
- a dilated pericentral region with radius $r = 1.15\, r_{\text{ref}}$ for all vessels located in the 25 percent of the relative distance to the central vein and reduced diameter outside of this region ($r = 0.9\, r_{\text{ref}}$).

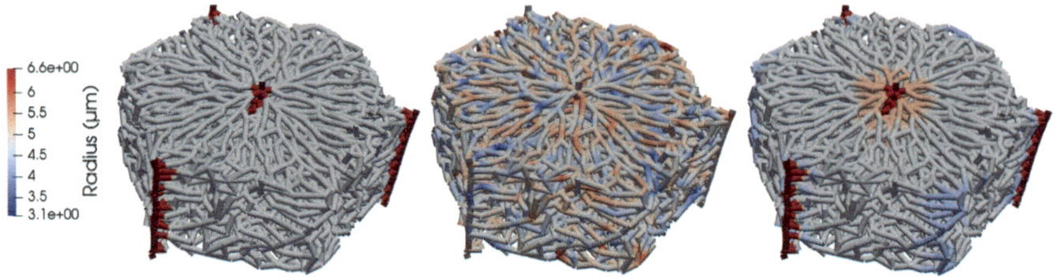

Fig. 5 The three considered lobule geometries: original lobule from [9] with sinusoidal radius as $r_{\mathrm{ref}} = 4.66\,\mu\mathrm{m}$ (left), Gaussian distribution of the vessels radii with mean of $4.85\,\mu\mathrm{m}$ and a standard deviation of $1.36\,\mu\mathrm{m}$ (center), dilated pericentral region with radius $r = 1.15\,r_{\mathrm{ref}}$ for all vessels located in the 25% of the relative distance to the central vein and reduced diameter outside of this region ($r = 0.9\,r_{\mathrm{ref}}$) (right). The same color-map from the normal distribution (without the lowest maximum of all three geometries) was set for all geometries for better visualization purpose

For each geometrical distribution, both flow and transport were computed. The results of the computations are summarized in Figs. 6, 7, 8, and Table 2.

In Fig. 6 we see that the overall aspect of the flow-related results is similar for all three distributions, which is to be expected since the same total flow was imposed. However, the distribution with a dilated pericentral region leads to slightly lower velocities around the central vein, and the Gaussian distribution leads to a flow field that is at some points different from the reference flow field.

The differences between the three distributions can also be seen from the pressure plots. The most striking difference is between the reference distribution (Fig. 6g) and the one dilated at the pericentral region (Fig. 6i), where the pressure drop from the PV value towards the CV value is much more pronounced and occurs much closer to the portal veins. This prompts us to believe that there is a systematic relationship between the dilation of the pericentral region and the pressure field across the lobule. To confirm this hypothesis, further simulations would need to be done to obtain a statistically relevant insight into this tendency.

As expected from the properties of Eq. (29), the concentration maps in Fig. 8 follow the velocity field, so the observations are analogous: comparable overall, with slight change in concentration near the central vein for the dilated pericentral distribution and local variations in concentration for the Gaussian distribution.

Common to all three distributions, we observe a delay effect from concentration maps in Fig. 8a–c to those in Fig. 8d–f. At the peak of the input concentration profile ($t_{\mathrm{P}} = 11$ s, Fig. 8a–c), the concentration is considerably higher in the regions close to the three portal veins. However, at 50% of the peak ($t_{1/2} = 43$ s, Fig. 8d–f), the compound concentration is higher in the regions further from the portal veins. This is connected to the properties of the flow field apparent from Fig. 6, but more importantly is an effect that can be further studied both via numerical simulations and via imaging techniques like those mentioned in Sect. 3.1.

An additional aspect amenable for distribution comparison is the CV concentration profile, shown in Fig. 7, and the mean transit time (MTT) computed from

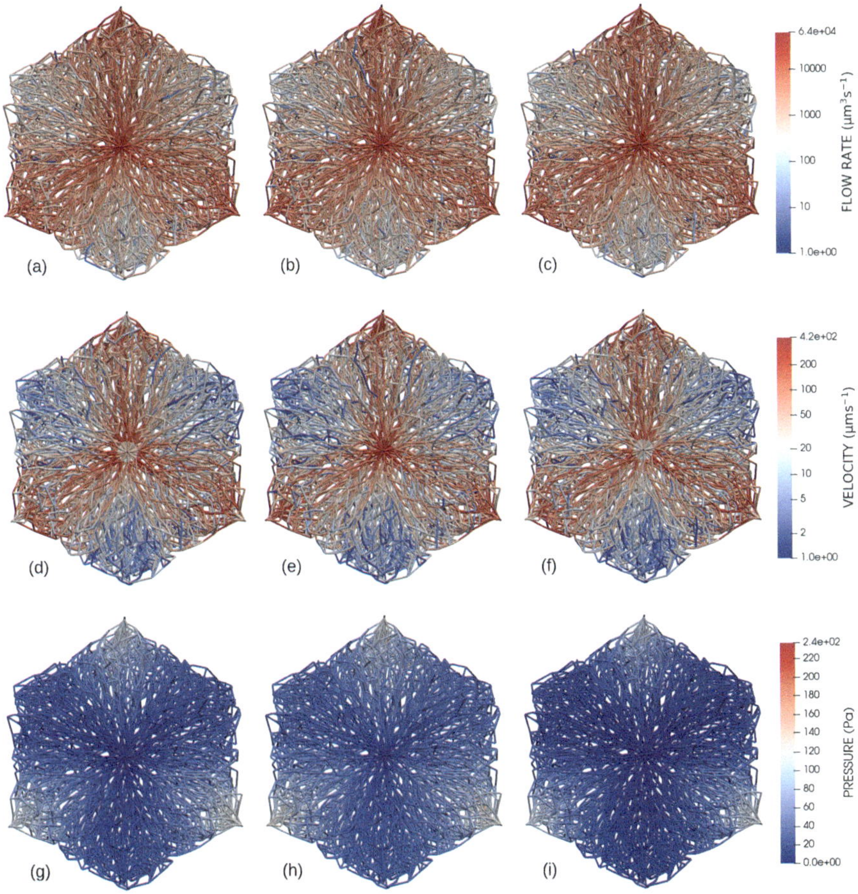

Fig. 6 Flow-related quantities for three distributions of vessel radii—reference (left: flow rate in (**a**), velocity in (**d**), pressure in (**g**)), gaussian (center: flow rate in (**b**), velocity in (**e**), pressure in (**h**)), and dilated at the pericentral region (right: flow rate in (**c**), velocity in (**f**), pressure in (**i**)). We can see that while the results for all three distributions look quite similar, the distribution with a dilated pericentral region leads to slightly lower velocities around the central vein, and the Gaussian distribution leads to a flow field that is at some points different from the reference flow field. There is also a hint of systematic tendency in the comparison of pressure profiles for the reference (**g**) and dilated (**i**) distributions. For the dilated distribution, the pressure drop is more significant and occurs closer to the portal veins

Eq. (27) listed in Table 2. From the latter, we see that the MTT is approximately 10% higher for the dilated pericentral distribution. For the Gaussian distribution the MTT is approximately equal to that of the reference configuration. This suggests that uniform changes to the network structure do not have much influence on MTT.

Overall, we have demonstrated that even in the simple scenario with pure advection inside the lobule and no exchange with hepatocytes, our model of flow and transport can capture slight variations in properties of the lobule vascular network

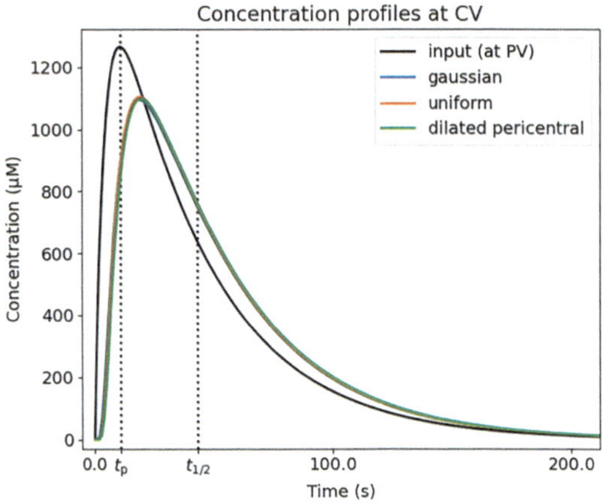

Fig. 7 Concentration profiles at the output (CV) for the three vessel radius distributions compared to the imposed input (PV) concentration profile. For the peak of input profile ($t_{\mathrm{P}} = 11$ s) and 50% of the peak value ($t_{1/2} = 43$ s), full concentration maps across the lobule are shown in Fig. 8. The main differences between the output concentration profiles are represented by MTTs computed in Table 2

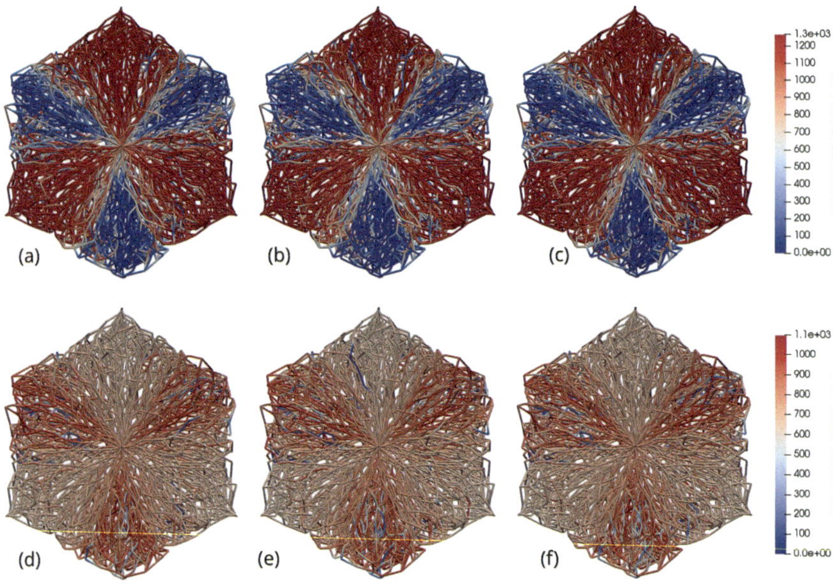

Fig. 8 Concentration maps across the lobule at input profile peak ((**a**), (**b**), (**c**)) corresponding to $t_{\mathrm{P}} = 11$ s, and at 50% of peak value ((**d**), (**e**), (**f**)) corresponding to $t_{1/2} = 43$ s for different distributions of vessel radii—reference (left), Gaussian distribution (center) and dilated at the pericentral region (right). Since for this case we are modelling pure advection without exchange with surrounding cells, the concentration maps are directly related to corresponding velocity fields (cf. Fig. 6), and thus the observations on differences between radius distributions are similar to those for flow data. The concentration maps are quite similar overall, the main differences being in specific places inside the lobule, e. g., slight change in concentration near the central vein for the dilated pericentral distribution, or the variation in concentration in particular places for the Gaussian distribution. Common to all distributions, there is a delay effect between t_{P} and $t_{1/2}$, where the concentration is still increasing in regions not directly connected to the portal veins when the concentration in the regions directly connected to the portal veins is already decreasing

Table 2 MTT comparison for different distributions of vessel radii

Radius distribution	MTT (s)
Uniform	8.012
Gaussian	8.011
Dilated pericentral	8.910

and measure their impact both on the level of micro-architecture and higher level indicators like MTT. This is of significant value, e. g., for pathologies connected to slight constriction or dilation of capillaries inside liver lobules. For more extensive simulations, see [2]. We have seen that, for a full understanding of this type of disease, a multi-level approach is necessary. This idea will be further developed in the next section.

3.3 *Case 3: Toxicity*

Prediction of toxicity *in vivo* based on *in vitro* experiments is a central question in the context of drug development [3]. In [5] we presented gradually more complex modeling strategies in order to evaluate under which conditions extrapolation to *in vivo* toxicity is feasible. The study was done in the context of paracetamol (APAP) toxicity which is known to kill hepatocytes at high dose [13]. In the *in vivo* experiments, the drug was injected in the peritoneum of mice and then transported to the liver where the hepatocytes eliminate it. If the dose is too high, another elimination pathway is triggered and a toxic compound is formed which will trigger a cascade of events leading eventually to the cell death.

3.3.1 Specific Equations and Boundary Conditions

The most sophisticated model developed in [5] involves a multi-compartment approach with the representation of a liver lobule (the smallest functional unit of the liver) with blood vessels and hepatocytes (Fig. 9).

This model reproduces the key events that happen during APAP elimination. APAP is injected in an homogeneous peritoneum compartment with linear elimination to a homogeneous systemic blood compartment. This compartment is then coupled to a liver lobule by prescribing the same concentration at the inlet of the lobule. The concentration at the liver outlet is then reinjected into the systemic blood compartment. A linear elimination term in the systemic blood compartment

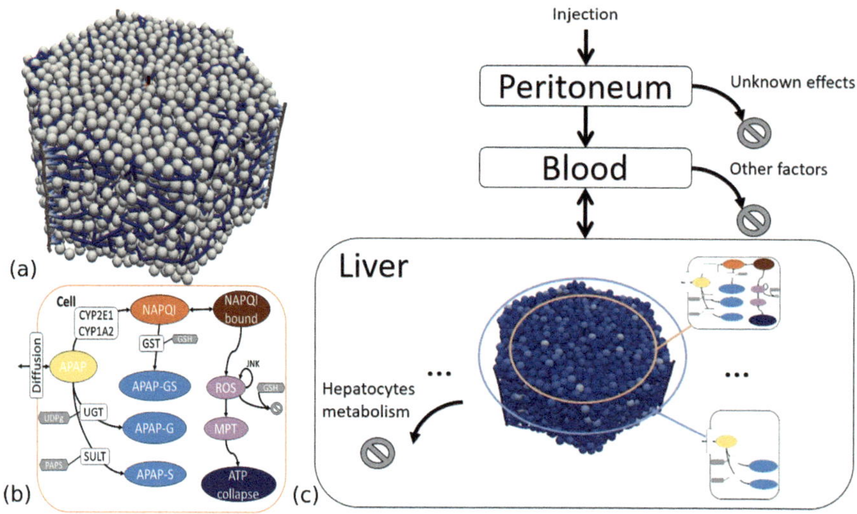

Fig. 9 Multi-scale toxicology model: (**a**) liver lobule with hepatocytes in contact with blood vessels, (**b**) metabolic model of APAP elimination, (**c**) full multiscale model coupled to external compartments [5]

accounting for elimination by other means than hepatocytes is also considered. This translates to:

$$\frac{\mathrm{d}c^{pe}}{\mathrm{d}t} = -Fk^{pe}c^{pe}, \tag{30}$$

$$\frac{\mathrm{d}c^{b}}{\mathrm{d}t} = Fk^{pe}c^{pe}\frac{V^{pe}}{V^{b}} + \frac{Q_{\text{liver}}}{V^{b}}\left(c^{p}(\text{outlet}) - c^{b}\right) - k^{\text{others}}c^{b}, \tag{31}$$

where c^{pe} is the concentration in the peritoneum compartment, k^{pe} is the linear elimination constant, V^{pe} is the volume of the peritoneum compartment, V^{b} is the volume of the systemic blood compartment, F is the dose dependent bio-availability fraction, Q_{liver} is the total liver flow rate, $c^{p}(\text{outlet})$ is the concentration leaving the liver lobule through blood at CV, c^{b} is the concentration entering the liver lobule through blood at PVs, and k^{others} is the linear elimination constant from other effects beside hepatocytes.

Blood flow modeling in the liver lobule is performed as described in Sect. 2.5.2. Regarding transport of APAP, diffusion in the blood vasculature is neglected and an exchange term accounting for APAP transport to the hepatocytes via passive diffusion is introduced. Equation (24) thus can be written as

$$\frac{\partial}{\partial t}c^{p} + \nabla \cdot (c^{p}\mathbf{v}) = -\sum_{h} \frac{PS^{ph}}{V^{p}}(c^{p} - c^{h})\delta(x^{h} - x), \tag{32}$$

where $c^p(x,t)$ here represents the APAP concentration in the blood vasculature at position x, c^h the APAP concentration in the hepatocyte at position x^h, connected to the vasculature at position x within the RVE at position x, P is the effective permeability and S^{ph} the contact surface between the hepatocyte h and its surrounding vascular volume, and V^p is the vascular volume associated with each RVE at position x. The sum goes over all connected hepatocytes in contact with the vascular volume at position x in the RVE (the delta function is non-zero if the hepatocyte is within the RVE of the considered vascular volume).

The APAP concentration inside each hepatocyte c^h is considered homogeneous, following the well-mixed assumption (see Sect. 2.4.3). Each hepatocyte of the liver lobule is a different compartment and has its own set of intra-cellular ordinary differential equations representing the APAP elimination metabolism. The parameters from one cell to another are different to account for cell to cell variability. The full equations are presented in [5], which we will not reproduce here. We will simply note that the coupling of the aforementioned advection-reaction equation to the hepatocytes' metabolism is achieved in each hepatocyte h as

$$\frac{\mathrm{d}c^h}{\mathrm{d}t} = \frac{PS^{ph}}{V^h}\left(c^p(x) - c^h\right)\delta(x^h - x) - \sum_j \nu_j^h \tag{33}$$

where c^p is the concentration of the vascular volume connected to the hepatocyte h, V^h is the latter volume, and ν_j^h are the reaction rates of the enzymes that trigger elimination of APAP in the cell h.

It is essentially a multiscale model that considers the whole body scale coupled to the lobule scale, and within the lobule, there are two "phases": blood and hepatocytes. But different hepatocytes act as different compartments.

Similarly to Sect. 3.2.2, when discretizing the equation for transport in the lobule vasculature, one arrives to a multi-compartment model, where each vascular small volume is a compartment interacting to its vascular neighbors and connected hepatocytes (Fig. 10).

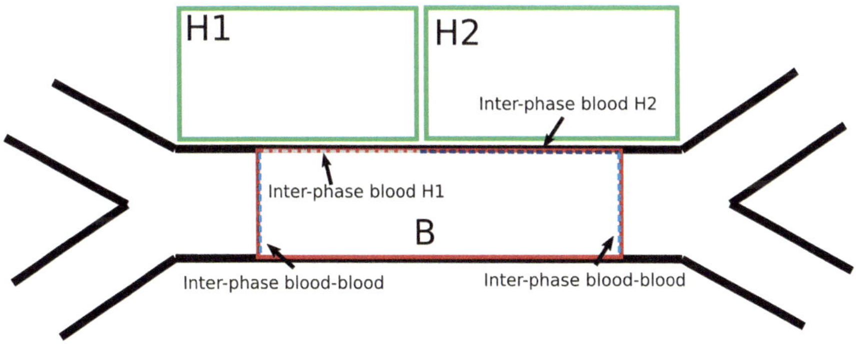

Fig. 10 Sketch of exchanges at discrete level of a single vascular volume (in red) and its interfaces with two hepatocytes (in green—H1 and H2) and surrounding blood volumes

The final model is thus a multiscale-model that ranges from the injection site to the individual cell level, which captures the key events of APAP transport and metabolism. It is important to note that in the model development process, it was shown that there could not be a one-to-one map between a space-free model and the multi-scale model presented here with spatial resolution of the liver lobule. This highlights that space representation matters to capture the correct kinetic of APAP concentration time profile and cell death.

3.3.2 Results

We investigate here the impact of the micro-architecture on APAP detoxification. The simulations were launched for a single 281 mg/kg dose and the same geometries as described in Sect. 3.2.2 (Fig. 5) are considered.

The results displayed on Fig. 11 show that the APAP detoxification multi-scale model is not sensitive to changes in the vascular architecture. The systemic blood concentration is however slightly higher in the reference case than in the dilated or Gaussian distribution cases. This is coherent with the fact that the reference case is associated with the highest cell death fraction (less cells to eliminate APAP implies higher concentration in the systemic blood compartment). This difference in cell death is small (less than 10 dead cells difference): the impact on the survival outcome of the total cell population of the liver lobule studied Fig. 11b is relatively small. At a given time point, the concentration values can be different between the different architectures (Fig. 11c–e and i–k), implying that the intra-lobule concentration' variations are different in each architecture but as stressed above, this does not impact the overall cell population. Although the alterations of the vascular architecture do not have a significant impact on the fraction of dead cells, around 3 h after injection (extrapolated as the final time of experiment), one can observe variations within the hepatocyte levels of ATP concentrations (Fig. 11f–h). This is particularly true for the case of Gaussian distribution of vessels (Fig. 11h). As a consequence the repeated exposure of this last architecture to a toxic dose would display a higher fraction of dead cells.

4 Discussion

We presented three models of micro-vascular transport in biological tissues that have been related to the same global framework. Each of them studies a different pathological condition related to blood microcirculation, and has been shown to reproduce or explain experimental or clinical observations. In their specific context, the model outputs generate measurable larger-scale outputs.

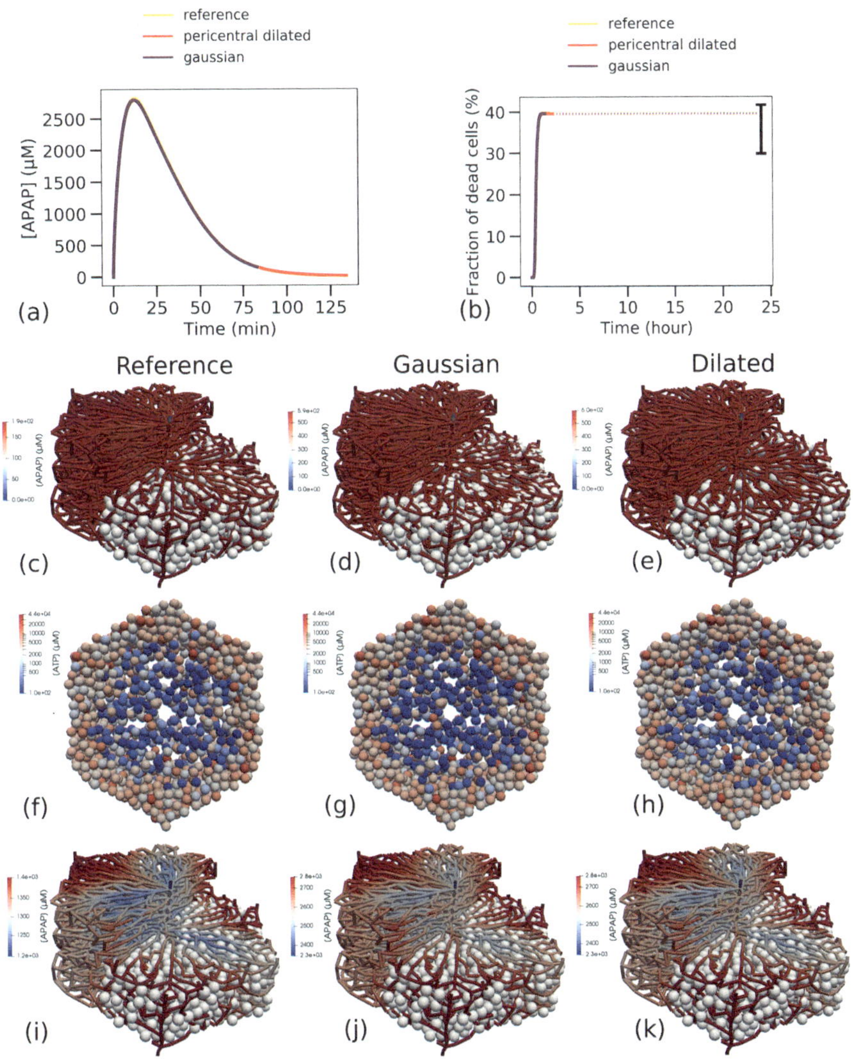

Fig. 11 Sensitivity of multi-scale model of APAP toxicity to micro-vasculature for a single 281 mg/kg dose: APAP concentration in the systemic blood compartment (**a**), fraction of dead cells over time with extrapolated fraction at 24 h (**b**), APAP concentration profile in the liver lobule one hour after injection for reference architecture (**c**), Gaussian distribution of vessels (**d**), dilated vessels (**e**), ATP concentration profile one hour after injection for reference architecture (**f**), Gaussian distribution of vessels (**g**), dilated vessels (**h**), APAP concentration 15 min after injection for reference architecture (**i**), Gaussian distribution of vessels (**j**), dilated vessels (**k**). For (**i, j, k**) the minimum concentration of the color map was set as 85% of the maximum concentration value independently for each architecture. For (**f, g, h**) the values are displayed in logarithm scale since they vary over several orders of magnitude

4.1 Strengths and Limitations of These Approaches

As presented above, the three models have the advantage of being geometry-independent, in the sense that the geometry on which the simulation is run can be easily adapted to other, healthy or pathological, situations. This facilitates the examination of diverse pathological conditions, in particular of micro-architectural distortions, on the blood flow and chemical species transport. Due to this aspect, the presented models can help to understand the repercussions of a micro-vascular disease while reducing the needed number of *in vivo* or *in vitro* experiments.
However, all three models have their specific limitations. The cases do not represent the vascular development of the studied pathology: as they take as a starting point the vascular geometry, there is no dynamic evolution of the tissue micro-architecture. Furthermore, the study is limited to a small range of microvascular changes, and their impact on hemodynamics and spread of a contrast agent or a single drug metabolism.

Due to their generality, the presented models can however readily be applied to other cases, and in this sense constitute *proof of concepts* [5, 21]. In particular, the generation of *in-silico* DCE-imaging data from a given, controlled, micro-architecture, constitutes the baseline to validate the inverse problem: to study how far compartment models are rich enough or limited to recover the correct tissue parameters from dynamic imaging data [21].

Inside the scope of this chapter, the mean transit time is the only marker of a transport behavior but does not encompass the complexity of the transport phenomena. Further studies have been lead on that topic in [21] and [2]. Although MTT is a common imaging biomarker [11], much more work is warranted on such global signal biomarkers to understand how sensitive they are to different microarchitectures.

4.2 Similarities and Differences in the Three Cases

In the three study cases, we investigate the impact of pathological alterations in the microvascular network's architecture. These geometry changes have an impact on *meso*- to *macro*-scale quantities such as blood flow and mean transit time (and going further with the number of dead hepatocytes in the third case). The different ways the disease impacts these quantities are summarized in Table 3.

In the first two test cases, geometry changes have a visible effect on pressure and blood flow. In both cases, the change in vessel radius can explain this change thanks to the relation between the resistance to flow and the radius of a blood vessel (see Eq. (26)). Unlike the lobule case, the vascularized tumor case explains the increase in diameter of the diseased region by introducing a higher micro-vascular density inside the tumor that remodels the upstream and downstream networks accordingly.

The mean transit time gives a first hint on the state of the microcirculation. Indeed, in the two first cases, the distorted architecture presents a significantly

Table 3 Summary of the influence of geometry on hemodynamics and transport

Case	Tumor	Lobule	Toxicity
Geometry changes	Increased micro-vascular density in the tumor region. Introduction of a necrosis region	Constant, Gaussian and peri-central dilated distributions of vessel radii	Constant, Gaussian and peri-central dilated distributions of vessel radii
Hemodynamics	The blood flow is higher in the tumoral region than in the healthy regions.	The pressure drop is located closer to the peri-central region in the random case and further for the dilated lobule	The pressure drop is located closer to the peri-central region in the random case and further for the dilated lobule
Transport	The plasma volume fraction is higher in the tumor region. The mean transit time is lower by around 15%.	The mean transit time increases by 11% in the dilated lobule compared to the reference and Gaussian cases.	No clear change

different mean transit time compared to the healthy reference geometry no matter the hemodynamic boundary conditions: the tumor case assumes a fixed pressure gradient which introduces an increased flow in the tumor, and even though in the lobule case the flow rate is prescribed at the portal triads, the mean transit time is higher in the distorted case compared to the healthy reference case. However, from a toxicity point of view, this increased mean transit time does not significantly affect the death of hepatocytes as the latter occurs over hours compared to seconds for the former.

5 Conclusion and Perspectives

This book chapter described mathematical models for blood flow and transport of injected species in the microcirculation, starting from first principles. Organ tissues are composed of multiple subsystems, blood vessels, interstitial space, cells, that can each receive, transform and/or transfer multiple components: their natural ones, mostly made of blood or water and naturally present, or injected molecules. Because these physiological processes occur partly at the cellular level in a heterogeneous manner, and there is a requirement to simulate these processes across large tissue areas, there exists a fundamental mathematical modeling challenge in reconciling these two scales.

Specifically, in this chapter, we presented a unified mathematical framework to model blood flow and transport (advection, exchange, diffusion, metabolism,...) of a medical imaging contrast agent in a vascularized tissue and drug toxicity in an organ functional unit (in the liver) as examples. We demonstrated on a few examples how the effect of different geometries, representing pathological or uncertainty in the normal vascular network morphologies, can be studied with such *in-silico* framework.

Acknowledgments The authors acknowledge the funding sources Funding by BMBF-grants LungSys (BMBF-FKZ 0316042B), LiSyM (BMBF-FKZ: 031L0045), MS-DILI (FKZ 031L0074F), LiSyM-Cancer (031L0257D), ANR projects iLite (ANR-16-RHUS-0005), and IFLOW (ANR-13-TECS-0006), EU project NOTOX and from the European Research Council (ERC) under the European Union's Horizon 2020 research and innovation programme (Grant agreement No. 864313).

Appendix

Compartment Models

Instead of considering REVs that include several phases, one can choose the simulation volumes along the phase borders, as for the microscale equations. In this case, every volume contains one phase only i.e., $\phi^p = 1$ for each volume considered, and the exchange of species occurs at the interface between the different considered volumes each containing one phase. The models are called compartment models. Compartments can for example be regions in one organ (e.g. blood, interstitial and cells) or entire organs. Consider for example an equation of a species diluted in a fixed volume of blood V. Then integrating the Eq. (24) on this volume gives

$$\int_V \frac{\partial c_i}{\partial t} dV + \int_V \nabla \cdot (\mathbf{u} c_i) dV = \int_V (\Re_i + \nabla \cdot (D \nabla c_i)) dV,$$

The $\Re_i$ term in the rhs. integral encompasses the reactions that the species i could undergo: $\Re_i = \sum_{j=1}^{r} \nu^{ij} \mathcal{J}_i^p / W_i$. It can be further re-written using the Gauss divergence theorem

$$\int_V \frac{\partial c_i}{\partial t} dV + \int_A \mathbf{u} c_i \cdot \mathbf{n} dA = \int_V \Re_i dV + \int_A D \frac{\partial c_i}{\partial \mathbf{n}} dA.$$

This is the integral form of the transport equation of species i in a single phase. By definition of the volume average quantities (noted with a bar), this equation becomes:

$$V \frac{\mathrm{d}\overline{c_i}}{\mathrm{d}t} = V \overline{\Re_i} + \int_A \left(D \frac{\partial c_i}{\partial \mathbf{n}} - c_i \mathbf{u} \cdot \mathbf{n} \right) dA.$$

(continued)

The compartments are often impermeable on the boundary except at specific locations where they can exchange with each other. Then the integral over the outer surface of the volume of the compartment p will have the form

$$\int_{A^p}\left(D\frac{\partial c_i^p}{\partial \mathbf{n}} - c_i^p \mathbf{u}\cdot\mathbf{n}\right) dA \approx \sum_{p'\neq p}\left(D^{pp'}A^{pp'}\frac{\partial c_i^p}{\partial n^{pp'}} - Q^{pp'}c_{i|p'}^p\right)$$

$A^{pp'}$ would be the exchange area at the interface between compartments p and p' and $Q^{pp'} = \int_{A^{pp'}} \mathbf{u}\cdot\mathbf{n}dA$ the blood flow rate through this exchange zone. The **well-mixed compartment** hypothesis [18] states that the concentration at the outlet borders of the compartment is the average concentration of the whole volume. This assumption can be summarized in our case as

$$\forall p' \neq p, c_{i|p'}^p = \begin{cases} \overline{c_i^p} & \text{if } Q^{pp'} \geq 0 \\ \overline{c_i^{p'}} & \text{otherwise} \end{cases}$$

Furthermore, the gradient of concentration across the border is assumed to depend on the two concentrations on each side of the border. Combining all assumptions leads to the following compartment kinetics:

$$V^p \frac{\mathrm{d}\overline{c_i^p}}{\mathrm{d}t} = V^p\overline{\Re_i^p} + \sum_{p'\neq p}\left(F(\overline{c_i^{p'}}, \overline{c_i^p}) - Q^{pp'}\overline{c_i^p}\right), \tag{34}$$

Which is an ordinary differential equation on the average concentration in the compartment $\overline{c_i^p}$.

References

1. Audebert, C., Bekheit, M., Bucur, P., Vibert, E., Vignon-Clementel, I.E.: Partial hepatectomy hemodynamics changes: experimental data explained by closed-loop lumped modeling. J. Biomech. **50**, 202–208 (2017)
2. Boissier, N., Drasdo, D., Vignon-Clementel, I.E.: Simulation of a detoxifying organ function: focus on hemodynamics modeling and convection-reaction numerical simulation in microcirculatory networks. Int. J. Numer. Meth. Biomed. Eng. **37**(2), e3422 (2021). https://doi.org/10.1002/cnm.3422. https://onlinelibrary.wiley.com/doi/10.1002/cnm.3422

3. Brecklinghaus, T., Albrecht, W., Duda, J., Kappenberg, F., Gründler, L., Edlund, K., Marchan, R., Ghallab, A., Cadenas, C., Rieck, A., Vartak, N., Tolosa, L., Castell, J.V., Gardner, I., Halilbasic, E., Trauner, M., Ullrich, A., Zeigerer, A., Demirci Turgunbayer, Z., Damm, G., Seehofer, D., Rahnenführer, J., Hengstler, J.G.: In vitro/in silico prediction of drug induced steatosis in relation to oral doses and blood concentrations by the Nile Red assay. Toxicol. Lett. **368**, 33–46 (2022). https://doi.org/10.1016/j.toxlet.2022.08.006. https://linkinghub.elsevier.com/retrieve/pii/S0378427422009869
4. DeLeve, L.D.: Vascular liver diseases. Curr. Gastroenterol. Rep. **5**(1), 63–70 (2003). https://doi.org/10.1007/s11894-003-0011-0
5. Dichamp, J., Cellière, G., Ghallab, A., Hassan, R., Boissier, N., Hofmann, U., Reinders, J., Sezgin, S., Zühlke, S., Hengstler, J.G., Drasdo, D.: In vitro to in vivo acetaminophen hepatotoxicity extrapolation using classical schemes, pharmacodynamic models and a multiscale spatial-temporal liver twin. Front. Bioeng. Biotechnol. **11**, 1049564 (2023). https://www.frontiersin.org/articles/10.3389/fbioe.2023.1049564
6. Formaggia, L., Quarteroni, A., Veneziani, A. (Eds.): Cardiovascular mathematics: modeling and simulation of the circulatory system. In: Modeling, Simulation & Applications, vol. 1. Springer, Milan (2009)
7. Furlan, A., Minervini, M.I., Borhani, A.A., Dioguardi Burgio, M., Tublin, M.E., Brancatelli, G.: Hepatic sinusoidal dilatation: a review of causes with imaging-pathologic correlation. Semin. Ultrasound CT MR **37**(6), 525–532 (2016). https://doi.org/10.1053/j.sult.2016.08.007
8. Gilles, R., Guinebretière, J.M., Lucidarme, O., Cluzel, P., Janaud, G., Finet, J.F., Tardivon, A., Masselot, J., Vanel, D.: Nonpalpable breast tumors: diagnosis with contrast-enhanced subtraction dynamic MR imaging. Radiology **191**(3), 625–631 (1994). https://doi.org/10.1148/radiology.191.3.8184038. PMID: 8184038
9. Hoehme, S., Brulport, M., Bauer, A., Bedawy, E., Schormann, W., Hermes, M., Puppe, V., Gebhardt, R., Zellmer, S., Schwarz, M., Bockamp, E., Timmel, T., Hengstler, J.G., Drasdo, D.: Prediction and validation of cell alignment along microvessels as order principle to restore tissue architecture in liver regeneration. Proc. Natl. Acad. Sci. U.S.A. **107**(23), 10371–10376 (2010). https://doi.org/10.1073/pnas.0909374107. https://pnas.org/doi/full/10.1073/pnas.0909374107
10. Hoffmann, U., Brix, G., Knopp, M.V., He, T., Lorenz, W.J.: Pharmacokinetic mapping of the breast: a new method for dynamic MR mammography. Magn. Reson. Med. **33**(4), 506–514 (1995). https://doi.org/10.1002/mrm.1910330408. https://onlinelibrary.wiley.com/doi/10.1002/mrm.1910330408
11. Lassau, N., Chami, L., Benatsou, B., Peronneau, P., Roche, A.: Dynamic contrast-enhanced ultrasonography (DCE-US) with quantification of tumor perfusion: a new diagnostic tool to evaluate the early effects of antiangiogenic treatment. Eur. Radiol. Suppl. **17**, 89–98 (2007)
12. Mauri, R.: Non-equilibrium thermodynamics in multiphase flows. In: Soft and Biological Matter. Springer, Dordrecht (2013). https://doi.org/10.1007/978-94-007-5461-4. http://link.springer.com/10.1007/978-94-007-5461-4
13. McGill, M.R., Jaeschke, H.: Metabolism and disposition of acetaminophen: recent advances in relation to hepatotoxicity and diagnosis. Pharm. Res. **30**(9), 2174–2187 (2013). https://doi.org/10.1007/s11095-013-1007-6. https://www.ncbi.nlm.nih.gov/pmc/articles/PMC3709007/
14. Meier, P., Zierler, K.L.: On the theory of the indicator-dilution method for measurement of blood flow and volume. J. Appl. Physiol. **6**, 731–744 (1954). https://doi.org/10.1152/jappl.1954.6.12.731. https://journals.physiology.org/doi/epdf/10.1152/jappl.1954.6.12.731
15. Parker, G.J., Roberts, C., Macdonald, A., Buonaccorsi, G.A., Cheung, S., Buckley, D.L., Jackson, A., Watson, Y., Davies, K., Jayson, G.C.: Experimentally-derived functional form for a population-averaged high-temporal-resolution arterial input function for dynamic contrast-enhanced MRI. Magn. Resonan. Med. **56**(5), 993–1000 (2006). https://doi.org/10.1002/mrm.21066. https://onlinelibrary.wiley.com/doi/abs/10.1002/mrm.21066
16. Pinder, G.F., Gray, W.G.: Essentials of Multiphase Flow and Transport in Porous Media. Wiley Online Library (2008)

17. Secomb, T.W., Pries, A.R.: Blood viscosity in microvessels: experiment and theory. Comptes Rendus Phys. **14**(6), 470–478 (2013). https://doi.org/10.1016/j.crhy.2013.04.002. https://linkinghub.elsevier.com/retrieve/pii/S1631070513000728
18. Sourbron, S.P., Buckley, D.L.: Classic models for dynamic contrast-enhanced MRI: classic models for DCE-MRI. NMR Biomed. **26**(8), 1004–1027 (2013). https://doi.org/10.1002/nbm.2940. https://onlinelibrary.wiley.com/doi/10.1002/nbm.2940
19. Tian, S., Xiong, Y., Liu, H., Niu, L., Guo, J., Liao, M., Xiao, S.Y.: Pathological study of the 2019 novel coronavirus disease (COVID-19) through postmortem core biopsies. Mod. Pathol. **33**(6), 1007–1014 (2020). https://doi.org/10.1038/s41379-020-0536-x
20. Ventimiglia, T., Linninger, A.A.: Mesh-free high-resolution simulation of cerebrocortical oxygen supply with fast fourier preconditioning. Int. J. Numer. Methods Biomed. Eng. **39**, e3735 (2023)
21. Vignon-Clementel, I.E., Jagiella, N., Dichamp, J., Kowalski, J., Lederle, W., Laue, H., Kiessling, F., Sedlaczek, O., Drasdo, D.: A proof-of-concept pipeline to guide evaluation of tumor tissue perfusion by dynamic contrast-agent imaging: direct simulation and inverse tracer-kinetic procedures. Front. Bioinformat. **3**, 977228 (2023). https://www.frontiersin.org/articles/10.3389/fbinf.2023.977228
22. Yin, Y., Sedlaczek, O., Müller, B., Warth, A., González-Vallinas, M., Lahrmann, B., Grabe, N., Kauczor, H.U., Breuhahn, K., Vignon-Clementel, I.E., Drasdo, D.: Tumor cell load and heterogeneity estimation from diffusion-weighted MRI calibrated with histological data: an example from lung cancer. IEEE Trans. Med. Imag. **37**(1), 35–46 (2018). https://doi.org/10.1109/TMI.2017.2698525
23. Yin, Y., Breuhahn, K., Kauczor, H.U., Sedlaczek, O., Vignon-Clementel, I.E., Drasdo, D.: Diffusion-weighted MRI-guided needle biopsies permit quantitative tumor heterogeneity assessment and cell load estimation (2021). https://doi.org/10.48550/arXiv.2103.00714. http://arxiv.org/abs/2103.00714. ArXiv:2103.00714 [physics]

Finite Element Software and Performance for Network Models with Multipliers

Cécile Daversin-Catty, Joseph P. Dean, and Marie E. Rognes

Abstract Hydraulic network models of flow in the microcirculation can give rise to mixed-domain partial differential equations; this is due to the regularity requirement on the flux and the 1D-0D coupling that emerges if Lagrange multipliers are introduced to impose bifurcation conditions. Although the finite element method is a natural approach to solving these problems, the implementation of mixed-domain finite element routines is generally non-trivial. The FEniCS and FEniCSx finite element libraries provide support for general mixed-domain finite element problems, and this chapter investigates their performance for hydraulic network models. We find that legacy FEniCS is limited to small networks due to the poor time complexity of the assembly routines. By contrast, the mixed-domain routines in FEniCSx have linear time complexity in the number of network branches, an important factor when solving large-scale hydraulic network problems.

Keywords Hydraulic networks · Finite element analysis · Performance

C. Daversin-Catty
Simula Research Laboratory, Oslo, Norway
e-mail: cecile@simula.no

J. P. Dean (✉)
University of Cambridge, Engineering Department, Cambridge, UK
e-mail: jpd62@cam.ac.uk

M. E. Rognes
Simula Research Laboratory, Oslo, Norway

Department of Mathematics, University of Bergen, Bergen, Norway
e-mail: meg@simula.no

A. Linninger et al. (eds.), *Quantitative Approaches to Microcirculation*,
SEMA SIMAI Springer Series 36, https://doi.org/10.1007/978-3-031-58519-7_4

1 Introduction

Many physical phenomena interact in the microcirculation, including blood flow and rheology in the vascular networks, solute transport to the surrounding tissue, and vessel wall mechanics. A diverse set of mathematical models of the microcirculation have been investigated over the past decades, see e.g. [4] and the references therein. One of the main challenges that remains is the multiple scales involved, including significant variations occurring in cells, in the surrounding tissue, and in whole-scale organs such as the brain, the heart, or the lungs. While 3D models can capture detailed dynamics and mechanisms, the prohibitive cost of such modeling approaches has led to a substantial interest in the development of more tractable reduced models. These include topologically one-dimensional models representing vascular or perivascular networks [12, 15, 34], as well as mixed-dimensional models coupling (1D) vascular networks and (3D) tissue [11, 13, 26–28, 30, 36]. Such 1D or 3D-1D models are also highly relevant in the geosciences [19, 22]. Mixed-dimensional coupling can also occur within networks, for instance, when boundary or bifurcation conditions are imposed through Lagrange multipliers. In this work, we consider hydraulic network models, a basic 1D-0D example of mixed-dimensional models with a diverse set of applications including the modeling of glymphatic and perivascular flow in the brain [42, 43].

The finite element method (FEM) is a natural approach to solving general mixed-domain partial differential equations (PDEs). However, implementing mixed-domain finite element routines can be non-trivial, especially in parallel; the complexity arises mainly from handling meshes of multiple domains of different topological dimensions, generating mixed-domain finite element kernels, and assembling the system of equations. Several finite element software packages have (varying levels of) support for these features [3, 5, 9, 10, 24, 35, 37], including the FEniCS project [2]. FEniCS is an open-source computing platform for solving partial differential equations. The FEniCS project started in 2003 and has since gathered a large community of users with a very wide range of applications. An automated framework [14] was developed in the core FEniCS libraries to address the challenges characterizing mixed-domain problems, relying on dedicated abstractions and algorithms. Despite the active development and maintenance over the years, some core attributes of FEniCS became obsolete and cumbersome to maintain, hindering the progress of the library. Limitations such as the lack of support for tensor-product elements, the challenges encountered when implementing operations that the software abstractions were not specifically designed for, and various performance-related issues often made implementing state-of-the-art methods complex, convoluted, and time-consuming. In 2018, FEniCS gave way to its successor FEniCSx [39–41], including major improvements over the legacy platform. Recent developments in the FEniCSx core libraries, introduced in [18] and inspired by [14], target the solution of general mixed-domain problems.

This chapter demonstrates how the mixed-domain framework of FEniCSx can be used to implement low-dimensional (1D-0D) hydraulic network models and investigates key aspects of their performance. Results are compared against a simulation code, based on [21], that is implemented using the legacy version of FEniCS.

2 A Minimal Mixed-Dimensional Model: Hydraulic Networks with Multipliers

2.1 Mathematical Model

We focus on a simple hydraulic network model of fluid flow in a capillary bed. Let Λ denote a tree-like network with n_e branches Λ_i such that $\Lambda = \cup_{i=1}^{n_e} \bar{\Lambda}_i$. Each branch corresponds to the centreline of a capillary and is associated with a unit tangent vector τ_i, which defines an arbitrary orientation [33]. The boundary of the network Λ is denoted by $\partial\Lambda$ and consists of inlet ($\partial\Lambda_{\text{in}}$) and outlet ($\partial\Lambda_{\text{out}}$) regions. The local inlet and outlet boundaries of each branch Λ_i are denoted by $\partial\Lambda_{i,\text{in}}$ and $\partial\Lambda_{i,\text{out}}$, respectively. The set of bifurcation points (i.e. the points at which the branches meet) is denoted by $\mathcal{B}$. Let the sets of indices of incoming and outgoing branches at b be denoted $I_{b,\text{in}}$ and $I_{b,\text{out}}$, respectively.

It is convenient to associate the hydraulic network with a directed graph G, composed of a set of n_e edges $\{e_i\}_{i=1}^{n_e}$ and n_v vertices $\{v_i\}_{i=1}^{n_v}$. Each topologically one-dimensional domain Λ_i is associated with an edge e_i whose direction is inferred from τ_i. Each topologically zero-dimensional boundary and bifurcation point is associated with a vertex v_i.

The pressure p_i of a fluid flowing in a cylindrical vessel Λ_i can be related to the flux through the vessel q_i by

$$\frac{\partial p_i}{\partial s} + \mathcal{R} q_i = 0 \text{ on } \Lambda_i, \tag{1}$$

where $\mathcal{R}$ is the flow resistance and s is the arc length along Λ_i. The pressure is assumed to be continuous over each bifurcation. The conservation of mass along each branch yields the condition

$$\frac{\partial q_i}{\partial s} = f \quad \text{on } \Lambda_i, \tag{2}$$

where f is a given source term. At the bifurcation points in the network, applying conservation of mass yields

$$[\![q]\!]_b = 0 \quad \forall b \in \mathcal{B}, \tag{3}$$

where the jump operator is defined as

$$[\![q]\!]_b := \sum_{i \in I_{b,\text{in}}} q_i|_b - \sum_{i \in I_{b,\text{out}}} q_i|_b, \tag{4}$$

i.e. the jump in the (discontinuous) field q at a bifurcation point b is the difference between the total incoming and outgoing flux at b.

2.2 A Mixed Finite Element Method

We consider a mixed finite element method to solve (1) with the conditions (2) and (3). Let the domain Λ be partitioned into a mesh $\mathcal{T}$ consisting of a finite number of 1D cells (intervals), and let $\mathcal{T}_i$ denote a conforming submesh containing only the intervals in $\mathcal{T}$ that make up Λ_i, where $1 \leqslant i \leqslant n_e$. The pressure finite element space, $P \subset L^2(\Lambda)$, is taken to be the space of discontinuous piecewise linear functions on $\mathcal{T}$. The cross-sectional flux q_i on each segment Λ_i belongs to the space $Q_i \subset H^1(\Lambda_i)$, defined as the space of continuous piecewise quadratics on $\mathcal{T}_i$. The bifurcation condition (3), which constrains the jump in the flux, is imposed through the Lagrange multipliers $\{\lambda_b\}_{b \in \mathcal{B}}$.

The weak form of the problem given by equations (1)–(3) then reads: find the fluxes $q_i \in Q_i$, pressure $p \in P$, and multipliers $\{\lambda_b\}_{b \in \mathcal{B}}$ such that

$$\sum_{i=1}^{n_e} \left(\int_{\Lambda_i} \mathcal{R} q_i v_i - \int_{\Lambda_i} p \frac{\partial v_i}{\partial s} \right) + \sum_{b \in B} [\![v]\!]_b \lambda_b = - \sum_{i=1}^{n_e} \int_{\partial \Lambda_i} p v_i, \tag{5}$$

$$\sum_{i=1}^{n_e} \int_{\Lambda_i} \frac{\partial q_i}{\partial s} \phi + \sum_{b \in B} [\![q]\!]_b \xi_b = \sum_{i=1}^{n_e} \int_{\Lambda_i} f \phi, \tag{6}$$

for all test functions $v_i \in Q_i$, $\phi \in P$, and $\{\xi_b\}_{b \in B}$.

3 Software Components and Implementation

In this section, we describe the software components used to implement a solver for the hydraulic network problem given in Sect. 2.2. We begin by giving a brief comparison of the mixed-domain functionality in legacy FEniCS and FEniCSx; for further details about the algorithms, software abstractions, and implementations, we refer the reader to [14] for legacy FEniCS and [18] for FEniCSx.

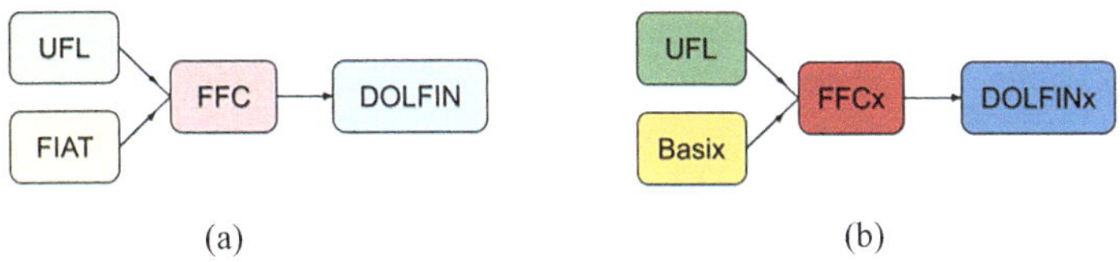

Fig. 1 The components making up FEniCS and FEniCSx. UFL is common to both, FIAT has been replaced by Basix in FEniCSx, and FFCx and DOLFINx are redesigned versions of FFC and DOLFIN, respectively. The arrows represent the (simplified) flow of information. (**a**) Legacy FEniCS. (**b**) FEniCSx

3.1 Mixed-Domain Abstractions and Algorithms in FEniCS and FEniCSx

FEniCS and its successor FEniCSx consist of a (partly overlapping) collection of software components [2] (cf. Fig. 1). An attractive feature of both FEniCS and FEniCSx is the high-level domain-specific Unified Form Language (UFL) that allows the weak form of a finite element problem to be defined using a notation that is close to the mathematical syntax [1]. UFL code is passed to the FEniCS (resp. FEniCSx) Form Compiler FFC [32] (resp. FFCx) which generates C++ (resp. C) finite element kernels to be used in the assembly process. The definition and tabulation of finite elements are handled by the FInite element Automatic Tabulator [25] (FIAT) in legacy FEniCS and by Basix [38] in FEniCSx. The problem-solving environment DOLFIN [31] and its successor DOLFINx provide tools for reading in meshes, creating function spaces, assembling the system of equations, applying boundary conditions, and writing the solution to file.

Native support for mixed-domain finite element methods was introduced in FEniCS in 2018, implementing the abstractions and algorithms described in [14]. To solve a mixed-domain problem, submeshes of each subdomain are created from an initial 'parent' mesh that conforms to all subdomain boundaries. Finite element function spaces can then be defined over each subdomain, and the matrices and vectors associated with forms containing test and trial functions from different domains can be assembled. This functionality can be used to solve a wide range of problems, including the hydraulic network model in Sect. 2.2, and we refer to [14] for several other examples, including a constrained Poisson problem, Stokes flow with non-standard boundary conditions, and a model for ionic electrodiffusion.

Mixed-domain functionality has recently been added to FEniCSx, taking advantage of the underlying upgrades in the library features and design. The mixed-domain algorithms in FEniCSx were introduced in [18] and are designed to be efficient, scalable, and parallel. The implementation is written in modern C++ (with an optional Python interface) and follows the functional, data-oriented design philosophy of DOLFINx, rather than the heavily object-orientated approach favoured by legacy DOLFIN. An advantage of the former is that it allows for greater flexibility and control in more complex applications, making it easier to implement schemes that do not necessarily fit within the existing abstractions. Where possible, data is structured in a way that aims to maximise spatial memory locality and scalable communication patterns are relied upon to improve performance in memory-bandwidth-limited applications. The reader is referred to [18] for further details, as well as parallel performance and scaling results for a range of applications including multiphysics problems, hybridised discontinuous Galerkin schemes, domain decomposition methods, and Lagrange multiplier problems.

3.1.1 Submeshes

In the hydraulic network problem in Sect. 2.2, each flux space Q_i is defined only over submesh $\mathcal{T}_i$ to obtain the desired regularity. FEniCS and FEniCSx both have the functionality to create these submeshes.

Legacy FEniCS implements a dedicated, lightweight `MeshView` class for representing and building the submeshes $\mathcal{T}_i$ as standard `Mesh` objects, as well as storing their relationship with the parent mesh $\mathcal{T}$. Submeshes can be created using the function `create` of the `MeshView` class, taking as input the subset of entities to be included in the submesh, defined by a marker (i.e. an array of integer tags associated with each entity in the parent mesh).

```
submesh = MeshView.create(marker, tag)
```

Maps from cells and vertices in the submesh to their corresponding entities in the parent mesh are automatically built and are accessible through the `MeshTopology` object of the generated submesh:

```
submesh.topology().mapping()[mesh.id()].vertex_map()
submesh.topology().mapping()[mesh.id()].cell_map()
```

The `MeshView` features are usable in parallel. The initial partitioning of the parent mesh is preserved i.e. each entity in the submesh is owned by the same process as in the parent mesh. While not optimal in terms of load balancing, this approach avoids the need for additional parallel communication that would be required when assembling coupling terms if the submeshes were repartitioned; for more details cf. [14].

FEniCSx implements a more explicit function, `create_submesh`, which, given a mesh (`mesh`) and a list of entities (`entities`) of dimension `dim`, returns a `Mesh` object containing only the entities in `entities`. It also returns maps relating cells, vertices, and geometry degrees-of-freedom in the submesh to entities, vertices, and geometry degrees-of-freedom in the original mesh, respectively.

```
submesh, entity_map, vertex_map, geom_map = create_submesh(
    mesh, dim, entities)
```

Since the submesh created is a `Mesh` object, any functionality in FEniCSx designed for `Mesh` objects, such as locating mesh entities, creating function spaces, file output, and even submesh creation, can be used. Unlike in legacy FEniCS, the FEniCSx implementation does not, in general, preserve the ownership of mesh vertices in submesh creation i.e. a vertex owned by a particular process in the submesh is not necessarily owned by the same process in the original mesh. This is to avoid corner cases that would require added parallel communication in various parts of the code, including in operations that would otherwise be local to a process (see [18] for more details).

3.1.2 Mixed-Domain Finite Element Kernel Generation

A finite element kernel computes the contribution of an element to the global system of equations. They are called in tight loops within a finite element code and therefore need to be optimised to achieve good performance in compute-bound applications. To avoid the user having to write optimised, problem-specific kernels by hand, both FEniCS and FEniCSx take advantage of automatic code-generation techniques.

In legacy FEniCS, given a finite element form specified with UFL, the form compiler FFC generates C++ code to tabulate the local finite element tensor. There are three supported integral types: cell integrals, exterior facet integrals, and interior facet integrals. In the homogeneous-dimension case, i.e. mixed-domain problems where the domains are of the same topological dimension, the kernel code generation procedure for cell and exterior facet (with respect to the parent mesh) integrals are handled using standard techniques in a similar manner to the single-domain case (see [32] for details). The extension to problems where the domains are of different topological dimensions is non-trivial, as detailed in [14], and relies on dedicated FFC abstractions. In the codimension-one case, i.e. when the domains differ in topological dimension by 1, integrals over the lower-dimensional domain are computed as surface integrals with respect to the higher-dimensional domain. This allows the existing exterior facet kernels to be repurposed to compute the mixed-dimensional coupling. However, the algorithm used involves combining two tensors and modifying certain degrees-of-freedom within each loop of the assembler (see [14, Sec. 6.2.2]), making the process complex and sub-optimal from a performance perspective.

The new version of the form compiler, FFCx, is simpler and more flexible than FFC; this is partly because Basix functions can be called from C++ at runtime, reducing the amount of generated code required [38]. The extra flexibility of FFCx allows for a simpler and more general implementation of mixed-domain kernel code generation, making a wider variety of coupling techniques possible (see [18] for details). The generated kernels are designed to be amenable to automatic vectorization, which many modern compilers use to improve performance. Code generation is optional in DOLFINx; for non-standard problems that do not fall into the abstractions provided by UFL, user-defined kernels can be written in C++ or Python (and compiled with Numba [29] for performance). It is also possible to call FFCx-generated kernels from Numba kernels. This allows user-defined kernels to be composed from generated kernels and custom operations.

3.1.3 Mixed-Domain Assembly

In a finite element problem, assembling the matrix or vector associated with a bilinear or linear form, respectively, typically consists of iterating over a subset of cells, computing the corresponding local tensor (i.e. the local matrix or vector for that cell), and inserting it into the global tensor A through a local-to-global degree-of-freedom map. The global tensors associated with mixed-domain forms are naturally block-structured, and the assembly can be performed block by block. While the assembly of the diagonal blocks is standard, since both test and trial functions belong to the same domain, the assembly of the off-diagonal blocks may be non-trivial. The relationship between the entities in each mesh is important for the mixed-domain assembly procedure.

In legacy FEniCS, in addition to the assembly routines for single-domain problems, there are dedicated assembly routines for mixed-domain problems. Assembly of the coupling (off-diagonal) blocks uses the maps discussed in Sect. 3.1.1 to relate entities in the integration domain mesh to entities in the meshes of other domains so that the correct degrees-of-freedom can be obtained [14]. All meshes in the form must have the same parent mesh. The homogeneous dimension case can then be handled using standard assembly techniques. In the codimension-one case, the assembly algorithm contains an additional inner loop to combine the tensors discussed in Sect. 3.1.2 (see [14, Alg. 2]).

The mixed-domain assembly routines are simpler in FEniCSx than in legacy FEniCS; the same assembly routines are used for both single- and mixed-domain problems, which is beneficial from a code maintenance perspective, and the codimension-one case is handled without adding an extra inner loop. The relationships between the meshes can also be more complex than in legacy FEniCS, since there is no need for all meshes to have the same parent, but maps relating entities in the integration domain mesh to entities in other meshes must be provided by the user; these can be created by composing the entity maps discussed in Sect. 3.1.1. Whilst this requires more input from the user compared to legacy FEniCS, it allows for greater control and flexibility in more complex applications, such as assembling forms involving nested submeshes (i.e. submeshes of submeshes).

3.2 Implementation of the Hydraulic Network Solver

We now describe how the mixed-domain functionality in FEniCSx can be used to implement the hydraulic network model in Sect. 2.2. We use the Python interface to FEniCSx, but the C++ interface could also have been used. The code is available in [16] and is based on [21].

We represent the directed graph G in code using the NetworkX Python package [23]. NetworkX provides data structures to represent standard graphs and tools to manipulate them. An illustration of a graph created with NetworkX is shown in Fig. 2a. A mesh of the network is created using a mesh generator, such as GMSH

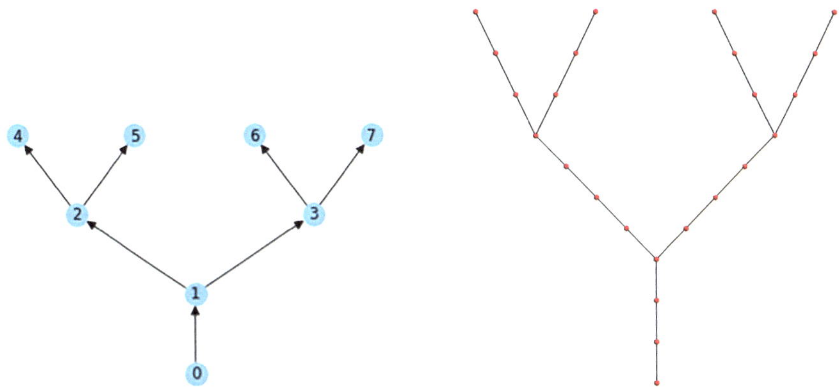

(a) (b)

Fig. 2 (**a**) The directed graph associated with a hydraulic network, created and represented using NetworkX [23]. (**b**) The network is discretized by a mesh of one-dimensional intervals, generated by GMSH [20]

[20], and consists of topologically one-dimensional intervals (Fig. 2b) embedded in $\mathbb{R}^2$ or $\mathbb{R}^3$. The vertices and edges of G are tagged to allow boundary and bifurcation conditions or heterogeneous material properties to be prescribed. Submeshes of the branches Λ_i are then created to allow the flux spaces Q_i to be defined. To reduce the number of submeshes required, we use NetworkX to create an edge colouring of G such that adjacent edges have different colours. Rather than creating a submesh for each edge, we create a submesh for each colour; this allows the desired regularity to be obtained whilst significantly reducing the total number of submeshes required for the networks of interest here. Next, the finite element function spaces are defined, and the forms in (5)–(6) are written using UFL. The linear system of equations can then be assembled and solved, and the solution can be written to file for postprocessing.

4 Results

The FEniCSx implementation of the hydraulic network model was verified by solving a simple problem with a known solution. In addition, the results of several benchmark problems were compared to results from a solver based on legacy FEniCS, described in [21]; both codes gave the same solutions. In this section, we compare the performance of the FEniCS and FEniCSx implementations.

4.1 Network Code Generation and Assembly in FEniCS vs FEniCSx

The performance tests were carried out on a laptop with an 8th generation Intel® Core™ i5-8350U CPU and 16 GB RAM. The legacy FEniCS simulations were run in a Docker container using the latest development version of FEniCS. The FEniCSx simulations were run in a Docker container with the latest version of the FEniCSx mixed-domain functionality installed.

We compare the form computation and assembly times on networks of sizes ranging from $n = 2$ generations (i.e. 3 edges) to $n = 6$ generations (i.e. 63 edges). We take $f = 0$ and prescribe a fixed pressure difference between the inlet and outlet boundaries. The computed fluxes and pressures are shown in Fig. 3.

Figure 4a shows the time taken to compile the finite element forms and assemble the system of equations for an initial execution of the solver, where the kernels have not previously been cached i.e. these timings include the code generation and compilation overhead. We do not show the time taken to solve the resulting linear system since both implementations use a sparse direct solver from UMFPACK [17] via PETSc [6–8], configured in the same way. The form generation and compilation times are approximately constant as the number of edges in the network increases for both FEniCS and FEniCSx, at least for the small networks considered here, with legacy FEniCS being slightly faster. The assembly time in FEniCSx is also approximately constant for these small networks, but we show in Sect. 4.2 that in larger problems the assembly time scales linearly with the number of edges. Assembling the system of equations is several orders of magnitude slower in legacy FEniCS; in these small networks, the majority of this time is spent in a code generation operation occurring within the assembler.

Figure 4b shows the form compilation and assembly times for a subsequent execution of the solver, where any generated code has been compiled previously and cached. The 'form compilation' time is now simply the time taken to fetch the kernels from the cache; it is therefore significantly reduced compared to Fig. 4a for both FEniCS and FEniCSx. In FEniCSx, the time to fetch the kernels is constant as the network size increases (at least in these small problems), whereas in legacy FEniCS it grows with n_e. Since no automatic code generation is carried out in the assembly process in FEniCSx, the assembly times are similar to those shown in Fig. 4a. The assembly time in legacy FEniCS scales as the cube of the number of network edges due to the poor time complexity of the mixed-domain assembly loop, which makes solving large or even moderate-scale network problems intractable. The cost of this operation was obscured by the code-generation operation occurring in the assembler in Fig. 4a due to the small networks considered here, but would, of course, dominate for larger n_e.

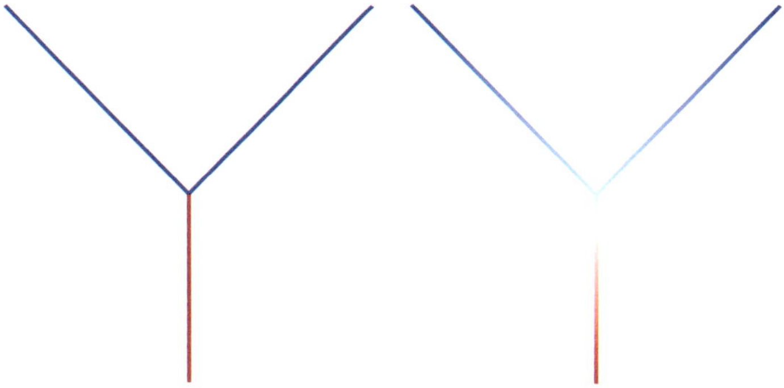

(a)

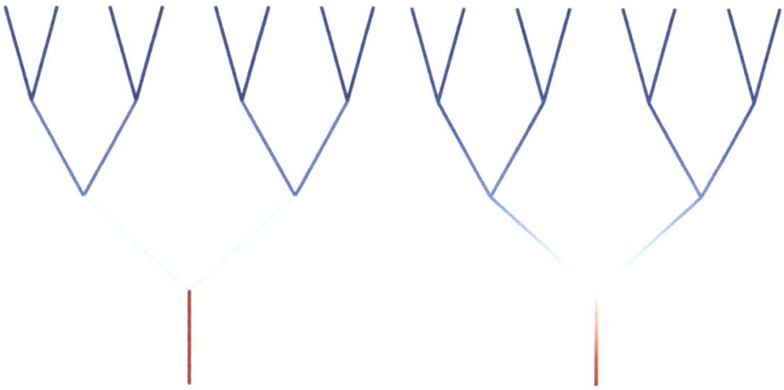

(b)

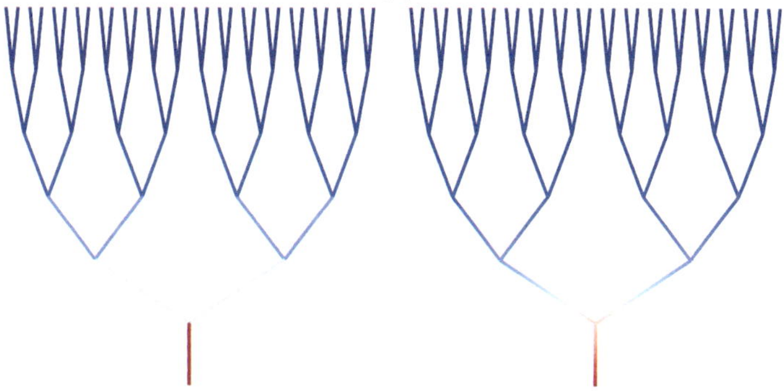

(c)

Fig. 3 Flux (left) and pressure (right) solutions of (5)–(6) for regular bifurcating networks with an increasing number of generations n. (**a**) $n = 2$ (3 edges). (**b**) $n = 4$ (15 edges). (**c**) $n = 6$ (63 edges)

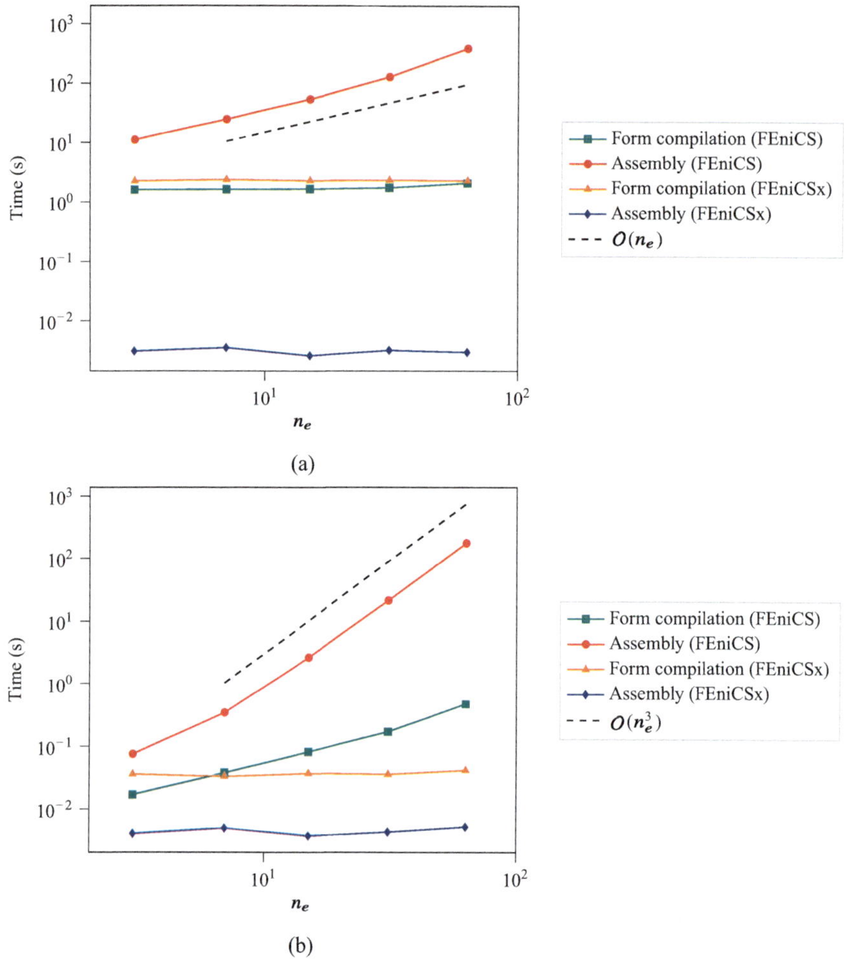

Fig. 4 Form compilation and assembly times against the number of network edges (n_e) for an initial execution of the solver (without cache) and a subsequent execution (with cache). (**a**) Without cache. (**b**) With cache

4.2 FEniCSx Performance on Larger Networks

We now investigate the performance of the mixed-domain algorithms in FEniCSx for larger hydraulic network problems. We consider the family of regular bifurcating networks in Fig. 3 from 2 generations (3 edges) to 24 generations (16, 777, 215 edges). The performance tests were run on an Apple MacBook Pro with an M2 Max CPU and 32 GB RAM. The time taken to create the submeshes, create the function spaces, assemble the matrix, and assemble the vector is illustrated in Fig. 5. The time taken by all operations scales linearly with the number of network edges.

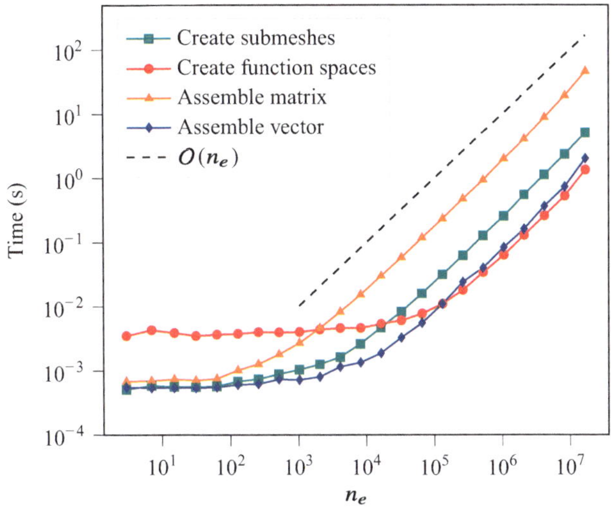

Fig. 5 The time taken to create the submeshes, create the function spaces, assemble the matrix, and assemble the vector as a function of network size in FEniCSx

5 Conclusions, Limitations, and Further Work

The mixed-dimensional framework [14] implemented in legacy FEniCS allows 1D hydraulic network models with bifurcation conditions imposed through Lagrange multipliers to be solved, as shown in [21]. However, due to several performance-limiting factors in legacy FEniCS, such as slow code generation and poor algorithmic complexity, it is not feasible to solve large-scale hydraulic network problems. The mixed-domain functionality in FEniCSx offers better performance and more flexibility for solving hydraulic network problems. The mixed-domain routines have linear complexity in the number of network edges, an important factor for solving large-scale hydraulic network problems. In addition, the mixed-domain framework in FEniCSx is still being actively developed.

Further work includes improving the handling of the Lagrange multipliers in the FEniCSx hydraulic network model. In legacy FEniCS, these are implemented using a dedicated function space with a single degree-of-freedom that can be used to represent a single Lagrange multiplier. These spaces are not yet supported in FEniCSx, so we have used a workaround that introduces more degrees-of-freedom than necessary. This is not optimal in terms of performance, but allows a fair comparison with the legacy version. Future work in FEniCSx will allow for a more natural and better-performing implementation of the bifurcation conditions.

Rheology in the vascular networks, solute transport, and microcirculation in general typically also involve the surrounding tissue. This can be modelled as a 3D domain embedding the 1D reduced network model. The mixed-dimensional coupling features of both legacy FEniCS and FEniCSx are, for now, limited to codimension-one couplings, i.e. 3D-2D, 2D-1D, or 1D-0D, as in the example presented in this chapter. The extension of the FEniCSx framework to 3D-1D coupling is currently in progress, entailing new features such as integration over edges and dedicated kernels for the mixed-dimensional local tensors.

Acknowledgments The authors graciously acknowledge Dr. Ingeborg G. Gjerde for the derivation and FEniCS-based implementation [21] of the hydraulic network model, and for the graph-colouring idea in Sect. 3.2. The authors also thank Dr. Jørgen S. Dokken for fruitful discussions on FEniCSx features and graph colouring. MER acknowledges support and funding from the Research Council of Norway (RCN) via FRIPRO grant agreement #324239 (EMIx). Support from the Engineering and Physical Sciences Research Council for JPD (EP/L015943/1 and EP/W026260/1) is gratefully acknowledged.

References

1. Alnæs, M.S., Logg, A., Ølgaard, K.B., Rognes, M.E.,Wells, G.N.: Unified form language: a domain-specific language for weak formulations of partial differential equations. ACM Trans. Math. Softw. **40**(2), 1–37 (2014)
2. Alnæs, M., Blechta, J., Hake, J., Johansson, A., Kehlet, B., Logg, A., Richardson, C., Ring, J., Rognes, M.E., Wells, G.N.: The fenics project version 1.5. Arch. Numer. Softw. **3**(100), 9–23 (2015)
3. Anderson, R., Andrej, J., Barker, A., Bramwell, J., Camier, J.-S., Cerveny, J., Dobrev, V., Dudouit, Y., Fisher, A., Kolev, T., Pazner, W., Stowell, M., Tomov, V., Akkerman, I., Dahm, J., Medina, D., Zampini, S.: MFEM: a modular finite element methods library. Comput. Math. Appl. **81**, 42–74 (2021)
4. Arciero, J.C., Causin, P., Malgaroli, F.: Mathematical methods for modeling the microcirculation. AIMS Biophys. **4**(3), 362–399 (2017)
5. Arndt, D., Bangerth, W., Feder, M., Fehling, M., Gassmöller, R., Heister, T., Heltai, L., Kronbichler, M., Maier, M., Munch, P., et al.: The deal. II library, version 9.4. J. Numer. Math. **30**(3), 231–246 (2022)
6. Balay, S., Gropp, W.D., McInnes, L.C., Smith, B.F. (1997). Efficient management of parallelism in object oriented numerical software libraries. In: Arge, E., Bruaset, A.M., Langtangen, H.P. (eds.) Modern Software Tools in Scientific Computing, pp. 163–202. Birkhäuser, Basel (1997)
7. Balay, S., Abhyankar, S., Adams, M.F., et al.: PETSc web page (2023). https://petsc.org/
8. Balay, S., Abhyankar, S., Adams, M.F., et al.: PETSc/TAO users manual. Technical Report ANL-21/39 - Revision 3.19, Argonne National Laboratory (2023)
9. Bastian, P., Blatt, M., Dedner, A., Dreier, N.-A., Engwer, C., Fritze, R., Gräser, C., Grüninger, C., Kempf, D., Klöfkorn, R., Ohlberger, M., Sander, O.: The dune framework: basic concepts and recent developments. Comput. Math. Appl. **81**, 75–112 (2021)
10. Blatt, M., Burchardt, A., Dedner, A., Engwer, C., Fahlke, J., Flemisch, B., Gersbacher, C., Gräser, C., Gruber, F., Grüninger, C., Kempf, D., Klöfkorn, R., Malkmus, T., Müthing, S., Nolte, M., Piatkowski, M., Sander, O.: The distributed and unified numerics environment, version 2.4. Arch. Numer. Softw. **4**(100), 13–29 (2016)
11. Caforio, F., Augustin, C.M., Alastruey, J., Gsell, M.A.F., Plank, G.: A coupling strategy for a first 3d-1d model of the cardiovascular system to study the effects of pulse wave propagation on cardiac function. Comput. Mech. **70**(4), 703–722 (2022)
12. Colebank, M.J., Qureshi, M.U., Olufsen, M.S.: Sensitivity analysis and uncertainty quantification of 1-d models of pulmonary hemodynamics in mice under control and hypertensive conditions. Int. J. Numer. Methods Biomed. Eng. **37**(11), e3242 (2021)

13. D'angelo, C., Quarteroni, A.: On the coupling of 1d and 3d diffusion-reaction equations: application to tissue perfusion problems. Math. Models Methods Appl. Sci. **18**(08), 1481–1504 (2008)
14. Daversin-Catty, C., Richardson, C.N., Ellingsrud, A.J., Rognes, M.E.: Abstractions and automated algorithms for mixed domain finite element methods. ACM Trans. Math. Softw. **47**(4), 1–36 (2021)
15. Daversin-Catty, C., Gjerde, I.G., Rognes, M.E.: Geometrically reduced modelling of pulsatile flow in perivascular networks. Front. Phys. **10**, 882260 (2022)
16. Daversin-Catty, C., Dean, J.P., Rognes, M.E.: FEniCSx hydraulic network solver source code (2023). https://github.com/cdaversin/networks_fenicsx. Accessed 26 Oct 2023
17. Davis, T.A. (2004). Algorithm 832: Umfpack v4.3—an unsymmetric-pattern multifrontal method. ACM Trans. Math. Softw. **30**(2), 196–199 (2004)
18. Dean, J.P.: Mathematical and computational aspects of solving mixed-domain problems using the finite element method. Ph.D. Thesis, University of Cambridge (2023)
19. Ferraz, P., Pereira, P., Abreu, E., Murad, M.A.: Recursive mixed multiscale model reduction for karst conduit-flow in carbonate reservoirs. Transp. Porous Media **139**, 527–558 (2021)
20. Geuzaine, C., Remacle, J.-F.: Gmsh: a 3-d finite element mesh generator with built-in pre-and post-processing facilities. Int. J. Numer. Methods Eng. **79**(11), 1309–1331 (2009)
21. Gjerde, I.G.: Graphnics: combining fenics and networkx to simulate flow in complex networks (2022). arXiv:2212.02916
22. Gjerde, I.G., Kumar, K., Nordbotten, J.M.: A singularity removal method for coupled 1d–3d flow models. Comput. Geosci. **24**, 443–457 (2020)
23. Hagberg, A., Swart, P., Chult, D.S.: Exploring network structure, dynamics, and function using networkx. Technical report, Los Alamos National Lab. (LANL), Los Alamos (2008)
24. Hecht, F.: New development in freefem++. J. Numer. Math. **20**(3–4), 251–266 (2012)
25. Kirby, R.C.: Algorithm 839: fiat, a new paradigm for computing finite element basis functions. ACM Trans. Math. Softw. **30**(4), 502–516 (2004)
26. Koch, T., Schneider, M., Helmig, R., Jenny, P.: Modeling tissue perfusion in terms of 1d-3d embedded mixed-dimension coupled problems with distributed sources. J. Comput. Phys. **410**, 109370 (2020)
27. Köppl, T., Vidotto, E., Wohlmuth, B., Zunino, P.: Mathematical modeling, analysis and numerical approximation of second-order elliptic problems with inclusions. Math. Models Methods Appl. Sci. **28**(05), 953–978 (2018)
28. Köppl, T., Vidotto, E., Wohlmuth, B.: A 3d-1d coupled blood flow and oxygen transport model to generate microvascular networks. Int. J. Numer. Methods Biomed. Eng. **36**(10), e3386 (2020)
29. Lam, S.K., Pitrou, A., Seibert, S.: Numba: a LLVM-based python JIT compiler. In: Proceedings of the Second Workshop on the LLVM Compiler Infrastructure in HPC, LLVM'15. Association for Computing Machinery (2015)
30. Laurino, F., Zunino, P.: Derivation and analysis of coupled pdes on manifolds with high dimensionality gap arising from topological model reduction. Math. Model. Numer. Anal. **53**(6), 2047–2080 (2019)
31. Logg, A., Wells, G.N.: Dolfin: automated finite element computing. ACM Trans. Math. Softw. **37**(2), 1–28 (2010)
32. Logg, A., Mardal, K.-A., Wells, G.: FFC: the fenics form compiler. In: Automated Solution of Differential Equations by the Finite Element Method: The FEniCS Book, pp. 227–238. Springer, Berlin (2012)
33. Notaro, D., Cattaneo, L., Formaggia, L., Scotti, A., Zunino, P.: A mixed finite element method for modeling the fluid exchange between microcirculation and tissue interstitium, pp. 3–25. Springer, Berlin (2016)
34. Olufsen, M.S., Peskin, C.S., Kim, W.Y., Pedersen, E.M., Nadim, A., Larsen, J.: Numerical simulation and experimental validation of blood flow in arteries with structured-tree outflow conditions. Ann. Biomed. Eng. **28**, 1281–1299 (2000)

35. Prud'Homme, C., Chabannes, V., Doyeux, V., Ismail, M., Samake, A., Pena, G.: Feel++: a computational framework for galerkin methods and advanced numerical methods. In: ESAIM: Proceedings, vol. 38, pp. 429–455. EDP Sciences, Les Ulis (2012)
36. Qohar, U.N.A., Munthe-Kaas, A.Z., Nordbotten, J.M., Hanson, E.A.: A nonlinear multi-scale model for blood circulation in a realistic vascular system. Royal Soc. Open Sci. **8**(12), 201949 (2021)
37. Schöberl, J.: C++ 11 implementation of finite elements in NGSolve. Technical Report ASC-2014-30, Institute for Analysis and Scientific Computing (2014)
38. Scroggs, M.W., Baratta, I.A., Richardson, C.N., Wells, G.N.: Basix: a runtime finite element basis evaluation library. J. Open Source Softw. **7**(73), 3982 (2022)
39. Scroggs, M.W., Dokken, J.S., Richardson, C.N., Wells, G.N.: Construction of arbitrary order finite element degree-of-freedom maps on polygonal and polyhedral cell meshes. ACM Trans. Math. Softw. **48**(2), 1–23 (2022)
40. The FEniCSx Developers: FEniCS project GitHub (2023). https://github.com/FEniCS. Accessed 25 Oct 2023
41. The FEniCSx Developers: FEniCS project website (2023). https://fenicsproject.org. Accessed 25 Oct 2023
42. Tithof, J., Kelley, D.H., Mestre, H., Nedergaard, M., Thomas, J.H.: Hydraulic resistance of periarterial spaces in the brain. Fluids Barriers CNS **16**, 1–13 (2019)
43. Tithof, J., Boster, K.A.S., Bork, P.A.R., Nedergaard, M., Thomas, J.H., Kelley, D.H.: A network model of glymphatic flow under different experimentally-motivated parametric scenarios. Iscience **25**(5), 104258 (2022)

A Fast-Fourier Preconditioned Schur Complement Method for the Simulation of Cerebrocortical Oxygen Supply

Thomas Ventimiglia and Andreas A. Linninger

Abstract The transfer of oxygen from blood to tissue involves mixed modes of transport between topologically distinct domains. To enable the high-resolution simulation of cortical oxygen supplied by an anatomically realistic microvascular network, we separate computational domains into a 1D graph representing the microvasculature and a rectangular array of 3D tissue voxels. Blood flow induces an orientation for which the vascular network is a directed acyclic graph, and this property enables efficient computation of RBC convection. Similarly, the 3D Cartesian array allows for extremely fast computation of oxygen diffusion with a fast-Fourier based Poisson equation solver. We find that the Schur complement method of domain decomposition decouples the tissue and vascular simulations without compromising the advantage of their respective fast solution methods, even with nonlinearity introduced by Michaelis-Menten reaction kinetics simulating metabolism in the tissue, and the Hill equation modeling dissociation of oxygen from RBCs into blood plasma before crossing the blood brain barrier. In particular, the discrete trigonometric transforms behind fast Poisson solvers efficiently precondition the Schur complement system for the tissue oxygen. With this method we are able to simulate oxygen supply both dynamically and at steady state for voxel resolutions approaching one micron side length without supercomputers.

Keywords Microvasculature · 1d-3d coupling · Domain decomposition · Fast solver · Cortical oxygen

T. Ventimiglia (✉)
University of Illinois at Chicago, Department of Biomedical Engineering, Chicago, IL, USA
e-mail: ventimi1@uic.edu

A. A. Linninger
University of Illinois at Chicago, Department of Biomedical Engineering and Department of Neurosurgery, Chicago, IL, USA
e-mail: linninge@uic.edu

A. Linninger et al. (eds.), *Quantitative Approaches to Microcirculation*,
SEMA SIMAI Springer Series 36, https://doi.org/10.1007/978-3-031-58519-7_5

1 Introduction

The availability of anatomically accurate 3D reconstructions of microvascular networks has fueled interest in computational modeling of oxygen transfer that is physiologically reliable down to smallest length scales [8, 10, 15, 17, 22]. The requirement of such a model to capture the multi-scale topological variability of vascular anatomical networks in the cerebral cortex, as well as the dense tortuous nature of the capillary bed, poses a massive computational challenge. For example, a cortical sample of $1 \times 1 \times 1\,\text{mm}^3$ at $1\,\mu\text{m}$ voxel resolution without any adaptive mesh refinement requires at least one billion equations.

Traditional finite element approaches as in [10] require a contiguously fitted mesh of sufficient complexity to resolve the capillary-tissue interface of the network region of interest. However, this method requires an excessively large number of elements [17]. Consequently, it becomes computationally intractable for simulating representative samples of the cerebral cortex ($> 0.5\,\text{mm}^3$) at high resolution.

The approach in this work follows the dual-mesh framework of Linninger et. al. in [22] and [15] to regard the coupled vascular and tissue domains as a 1D graph embedded into a course 3D mesh. This accommodates the extreme complexity of the capillary-tissue interface by requiring only a register of which polyhedral tissue elements contain vascular graph nodes in their interiors. These elements act as source terms for diffusion into the 3D domain with intensity governed by convection in the 1D graph: convection and diffusion are then solved simultaneously.

Previous massive simulations of oxygen extraction in the human and mouse cortex used unstructured tetrahedral meshes [15, 22]. Hartung et. al. in [17] substantially improved the computational efficiency and anatomical detail of such simulations by using a rectangular array of voxels to construct the tissue domain and the vascular network, with anatomically correct diameters and spatial positioning, in the same 3D space. Each voxel is representative of a Cartesian mesh cell and is registered as belonging to the vascular lumen or extravascular tissue according to its position relative to the vessel center-line. Tissue voxels adjacent to blood voxels or voxels large enough to encompass entire capillaries act as sources to supply oxygen to adjacent tissue cells. The granularity of the voxel mask can be tuned to desired microscale resolution, while interfaces at larger scales are continuously captured. The spatial separation circumvents excessively fine grain mesh detail at the smallest scales thus averting computational intractability. Moreover, the Cartesian framework does not need to store grid coordinates, thus it is termed *mesh-free*.

The current work builds upon the foundation set in [17] to further enlarge solvable problem sizes at substantially reduced computational cost as well as demonstrate the feasibility of dynamic simulation. This is in contrast to previous studies which only considered simulated oxygen fields at steady-state. This expansion is chiefly due to the separation of vascular and tissue domains by the Schur complement method [29], and the existence of fast solution algorithms for each individual domain. Use of a Cartesian voxel mesh for the tissue domain leads to a banded structured linear system for the discretized diffusion equations which are amenable to highly efficient preconditioning techniques using fast algorithms for discrete trigonometric transforms [14], such as those found in the open-source FFTW library [12]. Furthermore, when the vascular graph is given an orientation parallel to the blood flow it becomes a directed acyclic graph. It follows that the nodes of the graph can be ordered such that the linear system for convection is triangular, and thus easily invertible [3]. The inclusion of nonlinear reaction kinetics in the tissue and nonlinear mass transfer across the blood brain barrier only minimally affects the usefulness of these methods, requiring only the linearization of the core equations and subsequent implementation of the Newton-Raphson method to solve for each time-step.

Our method enables fully coupled nonlinear dynamic simulations of oxygen transport between blood and brain at high resolution, approaching the one micron length scale of more than one billion equations, on a standard desktop computer running MATLAB.

2 Models and Methods

Our method computes the steady-state oxygen field in both the vascular and tissue domains by solving a system of coupled nonlinear equations. Newton's method defers the solution to a set of linearized equations, and these are solved by decoupling the domains and solving the attendant Schur complement system. We will find that the Schur complement system can be solved efficiently by taking advantage of the topological structures of the two domains. The method can then be extended to the dynamic case. We provide an overview of the method and then go into each step in detail.

2.1 Overview

We use the finite volume method to solve for convective mass transport in the vascular domain and diffusive transport in the tissue domain at steady-state. The vascular domain is discretized into right circular cylinders with vascular segments as their centerlines, and the tissue domain is discretized into cubic voxels. For mass transfer between vascular and tissue domains we required oxygen bound to red blood cells to dissociate into the plasma before it is available to be transferred across the blood brain barrier. The concentration of free oxygen is related to the saturation percent of RBCs by the nonlinear Hill equation [18]. We also require oxygen in the tissue domain to be consumed with a metabolic rate following nonlinear Michaelis-Menten reaction kinetics [32]. The resulting set of nonlinear equations is solved by the Newton-Raphson method [27] whereby the equations are linearized and solved approximately at each Newton step. For an expanded discussion of the continuous problem and its discretization, see Appendix A.

Steady oxygen perfusion in Cartesian coordinates can be cast as a Poisson problem for which fast matrix decomposition algorithms are available [14]. *Fast* solvers exploit the fact that the eigen-decomposition of the finite volume discretization of the Poisson equation is analytically known, so instead of elimination steps, solutions can be directly evaluated using the *discrete sine transform (DST)*, which can be executed in $O(n \log n)$ operations, where n is the number of tissue elements [12]. Direct DST solution of the diffusion problem enables extremely efficient simulation of linear oxygen exchange even for problems exceeding one billion tissue elements. Since the eigenvalue decomposition and Poisson solution is analytical and direct, the explosion of the condition number hampering iterative algorithms at fine mesh resolutions does not occur [6].

The Poisson approach, however, does not apply to convective transport of blood oxygen. We therefore chose to incorporate oxygen supply to tissue by introducing source terms into the Poisson problem that are implicitly coupled to the blood oxygen tension. This approach, which eliminates unknown blood states, has the same effect as domain decomposition by forming a Schur complement of the linearized equations.

We find that the Schur complement system can be effectively preconditioned by solving a Poisson problem on the same rectangular domain, and the preconditioned system is then solved using the iterative linear algebraic system solver GMRES [30] to obtain a search direction for the next Newton iteration.

It is not necessary to form the Schur complement matrix itself, as GMRES only requires general linear operators. The output of the Schur complement operator requires the solution of an auxiliary linearized oxygen convection problem and since the vascular network is directed acyclic, this problem is triangular up to a relabelling of the graph nodes [3]. Hence the Schur complement operator can be evaluated cheaply at every step of GMRES.

This method to compute the steady-state oxygen field in the vascular and tissue domains can be extended to the dynamic case. Since the vascular and tissue domains evolve on different time scales, the differential equation system describing their simultaneous evolution is stiff; hence we use the backward-Euler method of time integration to ensure stability. This method requires the solution of a nonlinear system of equations at each step that differs only superficially from the equations at steady-state, so the preconditioned linearized Schur complement method works just as well for computing each time step [7].

2.2 *Graph Representation and Discretization of Vascular and Tissue Domains*

The center-lines of blood vessel segments are approximated as piece-wise linear arcs in 3D space. A vascular anatomical network (VAN) is thus made to correspond to a graph whose edges comprise the linear arc segments and whose nodes are their endpoints. The tissue domain, when regarded as a 3D rectangular array of voxels, similarly has a graph representation with nodes at the voxel centers and edges connecting each voxel to its at most six neighbors (cell-centered cubic finite volume elements): in other words, the tissue domain is represented as a 3D grid graph.

Each vascular edge has a constant radius associated with it so that it corresponds to a right circular cylinder in space with spherical end-caps: we call such a shape together with its interior a *capsule*. If a particular node has degree n, then there are n capsules meeting at the node. We may thus associate each node with the unions of halves of all n capsules that contain it in their intersection: see Fig. 1. This shape is the finite-volume element (control volume) associated with each vascular node.

The vascular and tissue graphs are united into a single graph by including edges connecting each vascular node to the tissue nodes closest to them. In particular, we include edges connecting each vascular node to the nodes of tissue voxels which intersect the surface of its control volume: Fig. 1 inset. See also Appendix 1 for more details. Nodes belonging to vessels with large radius relative to the voxel side length thus connect to multiple tissue nodes and supply them with oxygen.

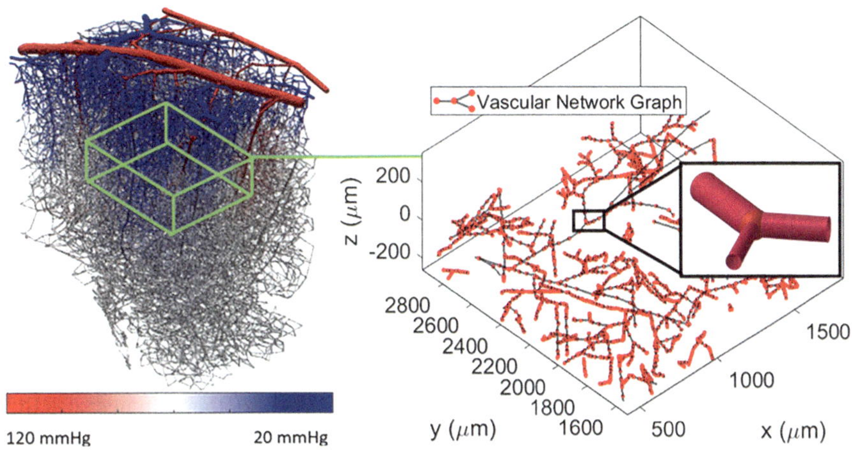

Fig. 1 Left: the vascular domain is a tubular network of blood vessels. Right: the centerline skeleton of is extracted as a spline network and the splines are discretized into a collection of line segments, forming the network graph. Inset: the control volume associated with a bifurcation node is the union of the interiors of three half capsules. Blood flows through the circular apertures at the midpoint of each connected edge and oxygen diffuses through the capsule surface

Conversely, nodes belonging to control volumes small enough to fit in a single voxel are associated with only that voxel. If a voxel contains many such capsules then they all supply it with oxygen.

In this way, the extent of the VAN geometry captured by the model is determined by the side-length of the voxels and the number of piece-wise linear arc segments. In the limit of small voxel side-lengths, the entire vascular network at all length scales (regarded as a collection of cylindrical capsules) is captured by the model. See [17] for details on the construction of the voxelized network and tissue complex.

2.3 Simulation of Blood Flow in the Vascular Network

Blood is simulated as a biphasic suspension of red blood cells (RBCs) in fluid plasma. We approximate bulk blood flow in each edge by the Hagen-Poiseuille law for laminar fluid flow in a pipe. This relates volumetric flow to pressure drop across each edge, hence we assign prescribed pressure boundary values to designated inlet and outlet nodes. High pressure at the inlet is 120 mmHg and low pressure at the outlet is 5 mmHg. Resistance to blood flow is determined by variable viscosity, which is determined by edge diameter and hematocrit. A kinetic plasma skimming model determined the variable hematocrit throughout the network by enforcing splitting rules for RBCs at each bifurcation. More details on computation of the flow and RBC distribution can be found in [15]. See Appendix 3 for a discussion of the blood flow simulation in the dynamic case.

2.4 Oxygen Convection and Mass Transfer in the Vascular Domain

Each node of the vascular network is associated with a control volume comprised of the union of half-capsules meeting at the node. The net change of oxygen bound to RBCs in each control volume is given by the following system of ordinary differential equations, whose terms we will describe in detail:

$$V_{RBC}\frac{dc_v}{dt} = (M_{\text{out}} - M)c_v + M_{\text{in}}\overline{c}_v - \frac{\Gamma_v}{h_v}C_{2,v}^T S\left[C_{2,v}F_v(c_v) - C_{2,t}c_t\right]. \qquad (1)$$

V_{RBC} is the $n_p \times n_p$ diagonal matrix whose entries are the volumes of RBCs in each control volume and c_v is the unknown length n_p vector of bound oxygen concentrations in each control volume.

2.4.1 The Connectivity Matrix of a Graph

The terms of this equation depend on the notion of the connectivity matrix of a graph C_1. If the vascular graph has n_f edges and n_p nodes, then the entries of its connectivity matrix $C_1 \in \{-1, 0, 1\}^{n_f \times n_p}$ are defined as $C_1(k, i) = 1$ if the directed edge k has node i as its initial node, -1 if i is its terminal node, and 0 otherwise. The transpose of this matrix is also called the incidence matrix of the graph.

2.4.2 The Convection Matrices

Let the operator $[\![A]\!] = \max(A, 0)$ set to zero all negative entries of the matrix A. The convection matrix M is defined as

$$M = C_1^T Q[\![C_1]\!] \tag{2}$$

where Q is the $n_f \times n_f$ diagonal matrix of positive RBC flow values (volume RBCs per unit time) on each edge. Each diagonal entry of Q is computed as $f_k H_k$ for volumetric blood flow f_k and hematocrit H_k in the edge k.

We also define the boundary convection matrices M_{in} and M_{out} as

$$M_{\text{in}} = D_{\text{in}} C_1^T Q[\![C_1]\!], \quad \text{and} \quad M_{\text{out}} = D_{\text{out}} C_1^T Q[\![-C_1]\!] \tag{3}$$

respectively. The decision matrices D_{in} and D_{out} are simply identity matrices of size $n_p \times n_p$ with all rows other that those corresponding the inlet (respectively outlet) nodes set to zero. As an operator these have the convenient effect of zeroing out all input rows not associated with a boundary node. Furthermore, the length n_p vector $\overline{c}_v$ contains prescribed boundary concentration values in the entries corresponding to the inlet nodes.

For a non-boundary node $i \in \{1, \ldots, n_p\}$ the ith entry of the vector Mc_v is the net flow of RBC bound oxygen through its associated control volume:

$$(Mc_v)_i = c_i \left(\sum_{k \mapsto (i,j)} f_k H_k \right) - \sum_{k \mapsto (j,i)} f_k H_k (c_v)_j. \tag{4}$$

Here, the symbol $k \mapsto (i, j)$ in the sum is meant to indicate that the edge $k \in \{1, \ldots, n_f\}$ is associated with the directed line segment connecting nodes i to j. Hence, the concentration of oxygen in each edge is given by the concentration in the upstream node, in keeping with the *upwinding* scheme of numerical fluid dynamics [26]. A similar analysis of the boundary convection matrices in (3) indicates that inlet nodes have a prescribed inflow of concentration given by the boundary values in the vector $\overline{c}_v$, and the outlet nodes have outflow equal to the flow value in their single incident edge.

It follows that the terms $(M_{\text{out}} - M)c_v + M_{\text{in}}\overline{c}_v$ in (1) represent net change of RBC bound oxygen within the network due to convection with the upwinding scheme. The remaining term then represents mass transfer into the tissue domain.

2.4.3 Fast Inversion of the Convection Matrices

At this point we make an important observation about the matrix $M_{\text{out}} - M$. The anatomical vascular networks we consider are directed acyclic graphs when the edges are oriented according to the direction of blood flow. This is reasonable, since we expect blood to flow smoothly from inlet to outlet without running into any circuits. Another way of saying this is that a directed graph is acyclic if and only if there exists a topological ordering of its nodes [3]. A topological ordering $<$ of the nodes $i, j \in \{1, \ldots, n_p\}$ satisfies the following property: $i < j$ if and only if j is downstream of i. That is, there is a directed path connecting i to j.

This means that the nodes of the vascular graph can be ordered such that $i < j$ for all j in the first term of (4) and $i > j$ for all j in the second term. It follows that there exists an ordering of the vascular nodes for which the matrix $M_{\text{out}} - M$ is *triangular* and hence easily inverted. We will assume from now on that a topological ordering has been chosen, as there exist numerous algorithms to find them.

2.4.4 Dissociation of Oxygen from RBCs to Plasma

Oxygen must become unbound from RBCs before it is available to cross the blood brain barrier. The Hill equation is an empirical relation between the saturation percent of RBCs and partial pressure of oxygen dissolved in plasma. If C_0 represents the concentration of oxygen in RBCs at 100% saturation and P_{50} is the oxygen tension at 50% saturation then the Hill equation is [13, 18]:

$$c_v = C_0 \frac{P^n}{P_{50}^n + P^n} \tag{5}$$

for oxygen tension P and an empirically determined number n called the Hill coefficient. Inverting this equation and multiplying by the effective solubility α of oxygen in plasma (Henry's constant) we obtain the nonlinear operator

$$F_v(c_v) = \alpha P_{50} \left(\frac{c_v}{C_0 - c_v} \right)^{1/n} \tag{6}$$

which relates the concentration of bound oxygen c_v to the concentration of free plasma oxygen $F(c_v)$. We may interpret F_v for a vector argument as applying the nonlinear function to each coordinate.

2.4.5 The Connectivity Matrix of the Domain Coupling

As discussed in Sec. 2.2, the vascular and tissue graphs are united by including edges connecting each vascular node to the tissue nodes whose voxels intersect the surface of the nodal control volume. Hence we are adjoining to the vascular and grid graphs a bipartite graph whose disjoint node sets are those of the vascular and grid nodes, and whose edges are the aforementioned coupling edges. The connectivity matrix of this bipartite graph furnishes the necessary tool to craft the equations describing mass transfer between domains.

Let the connectivity matrix of the bipartite coupling graph be the block matrix $C_2 = \begin{bmatrix} C_{2,v} & -C_{2,t} \end{bmatrix}$, where $C_{2,v} \in \{0,1\}^{n_c \times n_p}$ and $C_{2,t} \in \{0,1\}^{n_c \times n_t}$ are the blocks corresponding to the nodes connected to the vascular side (initial nodes) and tissue side (terminal nodes) of the bipartite graph respectively. Here, n_p is the number of vascular nodes, n_t is the number of tissue voxels, and n_c is the number of coupling edges.

When regarded as operators, $C_{2,v}$ and $C_{2,t}$ have the effect of projecting vectors of vascular and tissue values onto the interface between domains. The adjoint operators $C_{2,v}^T$ and $C_{2,t}^T$ then project them back onto their respective domains. It follows that the latter term of (1) represents the net diffusion of free plasma oxygen across the blood brain barrier, similarly to the approach of [21] where the mass transfer between domains is distributed over the surface of the cylindrical interface. Indeed, S is the size $n_p \times n_p$ diagonal matrix of surface area values across which oxygen can travel, Γ_v is the permeability of the vessel wall to oxygen in $\mu m^2/s$, h_v is the thickness of the vessel wall, here assumed to be a constant $1\,\mu m$, and c_t is the unknown vector of tissue oxygen concentrations. The surface areas in S are taken to be fractions of the control volume surfaces containing the vascular nodes. See Appendix 1 for an expanded discussion of the finite volume discretization of the vascular domain.

2.5 *Oxygen Metabolism and Diffusion in the Tissue Domain*

Each node of the grid graph is associated with a control volume equal to its containing voxel. Although voxels are classified as belonging to either blood or tissue, we will treat the blood voxels as equal to the tissue voxels in the simulation. This is to preserve the structure of the discretized diffusion problem and maintain the efficacy of the preconditioning method. This differs somewhat from the approach in [17], where the blood voxels are constrained to equal the values of the vascular nodes closest to them.

The net change of oxygen in each voxel is given by the following system of ordinary differential equations, whose terms we will describe in detail:

$$V_t \frac{dc_t}{dt} = -\Gamma_t h L c_t - \Gamma_t h L_{\text{bnd}} \overline{c}_t - V_t F_t(c_t) + \frac{\Gamma_v}{h_v} C_{2,t}^T S \left[C_{2,v} F(c_v) - C_{2,t} c_t \right]. \tag{7}$$

V_t is the $n_t \times n_t$ matrix of voxel control volumes and c_t is the unknown vector of tissue oxygen concentrations. Here we assume a constant voxel side length of h and volume of h^3; an alternative approach as in [15] would be to let V_t be only the fraction of voxel volume that lies in the extravascular space.

The last term of (7) is evidently representative of mass transfer between the vascular and tissue domains, as discussed in Sect. 2.4.5.

2.5.1 The Graph Laplacian

Diffusion of oxygen between neighboring voxels is proportional to the product of the gradient in concentrations and area through which the diffusion occurs [25]. Assuming a constant voxel side-length of h, it follows that the net diffusion through each voxel node is given by the discrete Laplacian of the grid graph. Indeed, if the connectivity matrix of the grid graph is given by C_3, then

$$\Gamma_t h \left(C_3^T C_3 c_t\right)_i = \Gamma_t h \sum_{\{i,j\}} \left[(c_t)_i - (c_t)_j\right] \tag{8}$$

for any interior voxel node $i \in \{1, \ldots, n_t\}$. The sum is over all *undirected* edges connecting node i to its adjacent nodes.

The matrix $C_3^T C_3$ is called the graph Laplacian of the grid graph because it is identical to the finite volume discretization of the continuous Laplacian Δ as seen in the Poisson equation $\Delta u = \phi$.

If c_t is the vector of only unknown interior voxels and $\overline{c}_t$ the vector of prescribed boundary values, then we may write the Laplacian in block form

$$C_3^T C_3 c_t = \ldots \begin{pmatrix} * & * \\ L_{\text{bnd}} & L \end{pmatrix} \begin{pmatrix} \overline{c}_t \\ c_t \end{pmatrix} \tag{9}$$

whence the first two terms in (7) represent the net change in tissue oxygen due to diffusion with its neighbors.

2.5.2 Fast Inversion of the Graph Laplacian with Dirichlet Boundary Conditions

The eigen-decomposition of the matrix $L = W^{-1} \Lambda W$ in (9) is analytically known. In fact, up to a constant scaling factor, the matrix by vector products Wv and $W^{-1}v$ can be computed by reshaping v into a 3D array with the same dimensions as the interior of the rectangular voxel domain and then taking its 3D discrete sine transform DST-I [11, 14]. The eigenvalues are also easily computed, since they comprise the values

$$\lambda_i + \lambda_j + \lambda_k \tag{10}$$

where

$$\lambda_i = 4\sin^2 \frac{i\pi}{2(n_x+1)}, \qquad i \in \{1, \ldots, n_x\} \tag{11}$$

$$\lambda_j = 4\sin^2 \frac{j\pi}{2(n_y+1)}, \qquad j \in \{1, \ldots, n_x\} \tag{12}$$

$$\lambda_k = 4\sin^2 \frac{k\pi}{2(n_z+1)}, \qquad k \in \{1, \ldots, n_x\}. \tag{13}$$

Here, n_x, n_y, and n_z are the number of interior voxels in the x, y, and z directions respectively.

It follows that to solve an equation of the form $Lu = v$ for u you need only take the 3D DST-I of v, divide by the eigenvalues, and then take the inverse 3D DST-I. The discrete sine transforms can be performed extremely quickly by using fast-transform algorithms.

Fast-transform based matrix decomposition algorithms exist for other common choices of boundary conditions as well, [14] but these were not implemented in this work.

2.5.3 Metabolism in the Tissue

Oxygen in the tissue is allowed to be metabolized over time following Michaelis-Menten kinetics [32]. The rate of oxygen consumption per unit volume is given by

$$F_t(c_t) = M_0 \frac{c_t}{\alpha P_0 + c_t} \tag{14}$$

where M_0 is the baseline oxygen demand per unit volume, P_0 is the oxygen tension at half consumption rate (the Michaelis constant) and α is the effective solubility of oxygen. We interpret F_t for a vector argument by applying the nonlinear function to each coordinate.

Evidently the term $V_t F_t(c_t)$ in (7) represents the net consumption of oxygen per unit time due to metabolic processes.

2.6 *Steady-State: Schur Complement of the Linearized System*

The master equations for the whole simulation are the coupled differential equations systems (1) and (7). We abbreviate them into block matrix equations as follows:

$$\begin{pmatrix} \dot{c}_v \\ \dot{c}_t \end{pmatrix} = \begin{pmatrix} A & E \\ G & B \end{pmatrix} \begin{pmatrix} c_v \\ c_t \end{pmatrix} + \begin{pmatrix} b_v \\ b_t \end{pmatrix} \tag{15}$$

where

$$A = V_{RBC}^{-1}\left[M_{\text{out}} - M - \tfrac{\Gamma_v}{h_v} C_{2,v}^T S C_{2,v} F_v\right] \tag{16}$$

$$E = \tfrac{\Gamma_v}{h_v} V_{RBC}^{-1} C_{2,v}^T S C_{2,t} \tag{17}$$

$$b_v = V_{RBC}^{-1} M_{\text{in}} \overline{c}_v \tag{18}$$

$$B = -F_t - V_t^{-1}\left[\Gamma_t h L + \tfrac{\Gamma_v}{h_v} C_{2,t}^T S C_{2,t}\right] \tag{19}$$

$$G = \tfrac{\Gamma_v}{h_v} V_t^{-1} C_{2,t}^T S C_{2,v} F_v \tag{20}$$

$$b_t = -V_t^{-1} L_{\text{bnd}} \overline{c}_t \tag{21}$$

2.6.1 The Newton Step

The operators A, G, and B contain the nonlinear functions F_v and F_t, so to solve the steady-state equation where the left side of (15) is zero we must set up the Newton-Raphson method and solve equations in the linearizations of the above operators.

Further abbreviate the vector of unknowns as $u = \begin{pmatrix} c_v \\ c_t \end{pmatrix}$ and the right side of (15) as the nonlinear function $f(u)$. The Newton-Raphson method produces a sequence of vectors $u^{(k)}$ for $k \geq 0$ approaching a solution to the steady-state equation $f(u) = 0$. After making an initial guess $u^{(0)}$, the subsequent iterates are defined by Bertsekas [4]

$$u^{(k+1)} = u^{(k)} - \alpha^{(k)} \left[J_f\left(u^{(k)}\right)\right]^{-1} f\left(u^{(k)}\right). \tag{22}$$

where $J_f = \partial f / \partial u$ is the Jacobian matrix of f and α is a scalar determined by approximate minimization of $f\left(u^{(k+1)}\right)$ with a quadratic line-search as described in Appendix C of reference [4].

The Jacobian matrix $J_f(x)$ is

$$J_f(x) = \begin{pmatrix} J_A(x) & E \\ J_G(x) & J_B(x). \end{pmatrix} \tag{23}$$

The Jacobian matrices of the individual nonlinear operators A, G, and B are identical to the forms given in Eqs. (16), (19), and (20) except replace every instance of F_v and F_t with the diagonal matrices $F_v'(x)$ and $F_t'(x)$ whose non-zero entries are the derivatives of F_v and F_t respectively evaluated at the coordinates of x.

$$F_v'(x) = \text{diag}\left(F_v'(x_1), \ldots, F_v'(x_{n_p})\right), \quad F_t'(x) = \text{diag}\left(F_t'(x_1), \ldots, F_t'(x_{n_t})\right) \tag{24}$$

where the scalar argument derivatives are

$$F_v'(y) = \frac{\alpha P_{50} C_0}{n y^{\frac{n-1}{n}} (C_0 - y)^{\frac{n+1}{n}}} \tag{25}$$

and

$$F_t'(y) = \frac{M_0 \alpha P_0}{(\alpha P_0 + y)^2} \tag{26}$$

2.6.2 The Schur Complement System

Each Newton step requires the solution of a linear equation of the form

$$J_f(x)u = v \tag{27}$$

or, upon separating again into the vascular and tissue components,

$$\begin{pmatrix} J_A & E \\ J_G & J_B \end{pmatrix} \begin{pmatrix} c_v \\ c_t \end{pmatrix} = \begin{pmatrix} v_v \\ v_t \end{pmatrix} \tag{28}$$

where we suppress the argument x of the Jacobian matrices for better readability.

We wish to solve for the vascular and tissue components separately so that we may take advantage of their respective fast algorithms. Using block Gaussian elimination to solve for c_v in the first row, we obtain the following equivalent system of matrix equations:

$$\left(J_B - J_G J_A^{-1} J_E\right) c_t = v_t - J_G J_A^{-1} v_v \tag{29}$$

$$c_v = J_A^{-1} (v_v - E c_t) \tag{30}$$

Equation (29) is the *Schur complement system* whose solution gives us the steady-state tissue oxygen values. If we can solve the Schur complement system for c_t, then we use (30) to obtain the corresponding steady-state vascular oxygen values.

The computation of c_v from (30) as well as the computation of the Schur complement matrix require the solution of equations of the form $J_A u = v$. After observing in Sect. 2.4.3 that $M_{\text{out}} - M$ is triangular, it is not hard to see that the matrix J_A is also triangular; hence, the matrix by vector products $J_A^{-1} v$ can be computed speedily.

Furthermore, if we use a Krylov subspace iterative method like GMRES to compute c_t approximately, then we need not form the Schur matrix itself: we only need to implement it as a linear operator [29]. Once again, this only requires the computation of matrix by vector products $J_A^{-1}v$ and multiplication by other sparse matrices.

2.6.3 Preconditioning the Schur Complement System

To speed up the iterative solution of the Schur complement system even more, we precondition it by solving a Poisson problem on the same rectangular domain. In particular, we solve the system

$$P^{-1}\left(J_B - J_G J_A^{-1} J_E\right) c_t = P^{-1}\left(v_t - J_G J_A^{-1} v_v\right) \tag{31}$$

where $P = \Gamma_t h V_t^{-1} L$ is the scaled Laplacian block L from Eqs. (7) and (9).

This method of preconditioning is feasible since we can compute matrix by vector products $P^{-1}v$ using the fast matrix decomposition algorithm discussed in Sect. 2.5.2. Furthermore, there is precedent to the practice of preconditioning Schur complements by choosing a good approximation of their first term [1, 24] or by using sine transforms [9].

In this case, we argue that P is a good preconditioner for J_B since J_B differs from P only by the mass transfer terms on the main diagonal. More intuitively, since most of the computational domain is occupied by tissue voxels and the voxels occupied by the vascular network are relatively sparse (see column 4 of Table 3), we guess that diffusion processes dominate the equation for tissue steady-state and consequently, we can precondition the Schur complement system by solving a pure diffusion problem as a preprocessing step.

This intuition is borne out in Fig. 2 as we track the residual norm over many GMRES iterations for a modification of (31) that arbitrarily weakens the coupling term $J_G J_A^{-1} J_E$ by a factor $0 < \gamma < 1$. The preconditioning performs nearly as well in the case of weak coupling, indicating that P is nearly as effective of a preconditioner for J_B as it is for the whole Schur complement, and the influence of the coupling terms is relatively minor.

Note that the presence of reaction terms in J_B still keeps preconditioned GMRES from converging immediately: if both the reaction terms and coupling were absent, then (31) would be a pure diffusion problem and the preconditioning would solve it exactly in one step.

The result in Fig. 2 also shows that solving systems of the form $J_B x = b$ is highly non-trivial, especially when the tissue resolution is high and so the matrix is very large. This shows why we chose to formulate the Schur complement with respect to J_A instead of with respect to J_B; that is, we took the Schur complement to be $J_B - J_G J_A^{-1} J_E$ instead of $J_A - J_E J_B^{-1} J_G$. As discussed in Sect. 2.4.4, the network graph imbued with an orientation parallel to the flow field is a *directed acyclic graph*, and as a consequence the matrix J_A is triangular. Hence, systems of the form $J_A x = b$ can be solved cheaply by back-substitution, and so the Schur complement with respect to J_A can be implemented as an operator cheaply. The

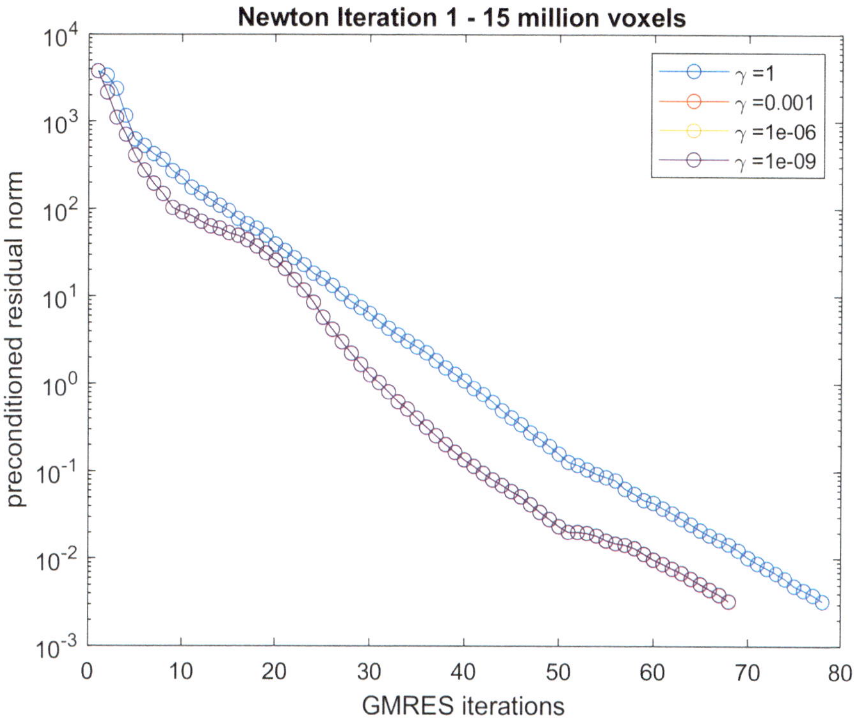

Fig. 2 Preconditioned GMRES is used to solve the Schur complement system involved in the first Newton step for ~15 million tissue voxels surrounding the ~134,000 segment anatomical network, with an additional parameter γ which modulates the strength of the coupling: $P^{-1}(J_B - \gamma J_G J_A^{-1} J_E)x = P^{-1}b$. The results for all $\gamma < 1$ lie on top of one another

alternative would require another layer of iterative solution to even evaluate the Schur complement.

2.7 *The Dynamic Case*

We now describe an implementation of the backward (implicit) Euler method to integrate the fully dynamic equation (15) in time. As in Sect. 2.6.1 we have the abbreviated equation $\dot{c} = f(c, t)$ where the time dependence comes from the variable RBC flow in the convection matrices. For initial guess $c^{(0)}$, we compute the next time step $c^{(\ell+1)}$ as [7]

$$c^{(\ell+1)} = c^{(\ell)} + \Delta t f\left(c^{(\ell+1)}, t^{(\ell+1)}\right) \tag{32}$$

where Δt is the fixed time step and $t^{(\ell+1)} = t^{(\ell)} + \Delta t$.

2.7.1 The Newton Step

At every time step we need to solve the nonlinear equation (32) for $c^{(\ell+1)}$ and, as in Sect. 2.6.1 we do this with the Newton-Raphson method. Set the current time step to equal the initial guess for Newton's method: $u^{(0)} = c^{(\ell)}$. Then, each Newton step produces an approximation to the next time step $u^{(k+1)} \approx c^{(\ell+1)}$ by the following equation

$$u^{(k+1)} = u^{(k)} - \alpha^{(k)} \left[I - \Delta t J_f \left(u^{(k)}, t^{(\ell+1)} \right) \right]^{-1} \left[u^{(k)} - c^{(\ell)} - \Delta t f \left(u^{(k)}, t^{(\ell+1)} \right) \right] \tag{33}$$

where I is an identity matrix, J_f is the same as in (23), and again $\alpha^{(\ell)}$ is a variable step-size which we find by a quadratic line-search algorithm.

2.7.2 The Preconditioned Schur Complement System

Every Newton step requires the solution of a linear system of the form

$$\left[I - \Delta t J_f \left(u^{(k)}, t^{(\ell+1)} \right) \right] u = v \tag{34}$$

As in Sect. 2.6.2 we may form the Schur complement system for the tissue states

$$S(x,t)c_t = b_s(x,t) \tag{35}$$

where, after suppressing the arguments for readability,

$$S(x,t) = I - \Delta t J_B - \Delta t^2 J_G \left[I - \Delta t J_A \right]^{-1} E \tag{36}$$

$$b_s(x,t) = \quad v_t + \Delta t J_G \left[I - \Delta t J_A \right]^{-1} b_v \tag{37}$$

and the vascular states are recovered as

$$c_v = (I - \Delta t J_A)^{-1} (v_v + \Delta t E c_t). \tag{38}$$

The matrix $I - \Delta t J_A$ differs from J_A only on the diagonal entries, so it follows that $I - \Delta t J_A$ is triangular as well as J_A, and so formation of the matrix by vector products $(I - \Delta t J_A)^{-1} v$ can be done speedily in the computation of b_s and $S(x,t)$ can be done speedily, and preconditioned GMRES remains a feasible strategy to obtain the candidate tissue states c_t.

To precondition the Schur complement system (36) we seek a good approximation of the matrix $I - \Delta t J_B$. Since this only differs from J_B by the diagonal entries, we can precondition it with the graph Laplacian L as well.

2.7.3 Summary of Method

We summarize the method for the fully coupled, nonlinear, dynamic simulation:

1. Compute the initial condition $c^{(0)}$ for the dynamic case by solving for the steady-state case.
   (a) Choose an initial guess $u^{(0)}$ for Newton's method. A good choice is to solve for the steady-state at low resolution and then interpolate up to higher resolution.
   (b) In each Newton step solve the preconditioned equation (31) with GMRES. In every step of GMRES compute the matrix by vector products $J_A^{-1}v$ in the Schur complement operator by solving a triangular system and compute the matrix by vector products $L^{-1}v$ with the fast-transform based matrix decomposition algorithm described in Sect. 2.5.2.
   (c) Use the computed tissue states c_t to recover the vascular states c_v with Eq. (30).
   (d) Compute $\alpha^{(k)}$ with a quadratic linesearch algorithm and use the computed values of c_v and c_t to compute the next Newton step $u^{(k+1)}$ via (22).
   (e) Repeat until convergence.
2. Set the initial guess $u^{(0)}$ for Newton's method to equal to the current time step $c^{(\ell)}$.
3. Solve the Schur complement equation (35) for the candidate tissue states using GMRES and the same fast-algorithms as for the steady-state solution.
4. Recover the vascular states using (38).
5. Compute $\alpha^{(k)}$ using a quadratic line-search algorithm and use the computed values of c_v and c_t to compute the next Newton step $u^{(k)}$ in (33).
6. Repeat Newton steps until convergence, and set the Newton solution to be the new time step $c^{(\ell+1)}$.

3 Application to Massive Steady-State Oxygen Exchange Simulation in the Cerebral Cortex

All numerical experiments were performed in MATLAB version R2022a on a desktop PC with an Intel Xeon W-2125 processor at 4.01 GHz and 256 GB of memory. For the fast-Poisson preconditioning we use the open-source FFTW library [12] to compute the discrete sine transform (DST-I) by way of a MATLAB front-end [33]. By way of comparison, we note that it was found in [17] that for the $3\times3\times1.3$ mm^3 network in Fig. 3 panel A and about 95 million tissue voxels with linear mass flux, the automatic solver [2] converged in about 19 h. With our improved methods, in this work we solved for 535 million tissue voxels with nonlinear mass flux on the same network and the same computer in one hour.

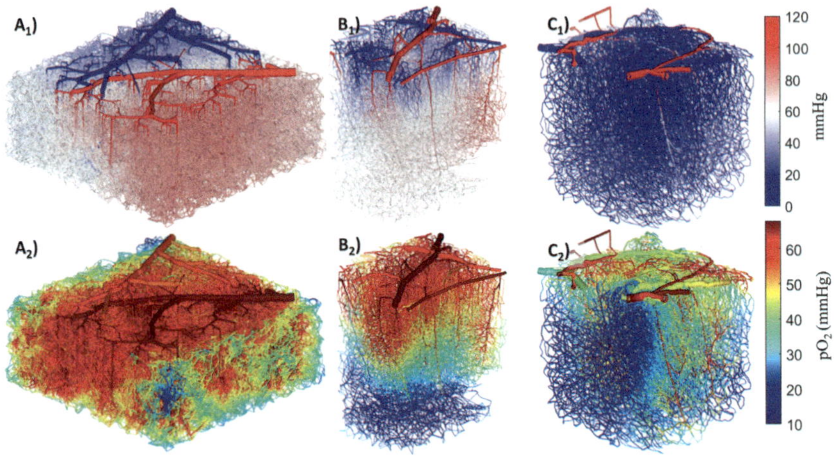

Fig. 3 Voxelized synthetic and realistic in-vivo networks. Top-row: blood pressure in mmHg. Bottom row: steady state blood oxygen partial pressure in mmHg. (**a**) $3 \times 3 \times 1.3$ mm^3 synthetic network [17] with 636,350 vessel segments. OEF = 12.3%. $CMRO_2$ = 9.8e-06 mol/L/s (**b**) 1.1 $\times 1.5 \times 1.8$ mm^3 in-vivo network [5] with 39,547 vessel segments. OEF = 10.1%, $CMRO_2$ = 3.2e–06 mol/L/s (**c**) $1.1 \times 1 \times 1.1$ mm^3 in-vivo network [5] with 134,229 vessel segments. OEF = 34.6%, $CMRO_2$ = 9.5e–06 mol/L/s

A thorough examination of theoretical convergence rates with respect to discretization length is outside the scope of this work.

A simple simulation in the case of a single capillary is shown in Appendix 3 as a demonstration of our methods viability in the dynamic case. Although the dynamic simulation is functional for all three of the test networks shown in Fig. 3, we refrain from putting them in the main body of the text because they have not been extensively validated yet.

3.1 Pure Diffusion of an Ideal Solute

To demonstrate the speed of the preconditioner described in Sect. 2.5.2 we solve the finite difference discretization of the 3D Poisson equation $Lu = v$ on an $n_x \times n_y \times n_z$ voxel rectangular domain with Dirichlet boundary conditions of zero and with a randomized right-hand side vector v. For all cases under one billion voxels, we set $n_x = n_y = 2^k - 1$, and $n_z = 2^{k-1} - 1$ for $k = 5$ through 10. Side lengths of this form are optimal for the fast transform algorithm [11]. Solution times are compared in Table 1 and Fig. 4 to different standard solvers. The results show that our method drives residuals to arbitrarily small values and can solve the Poisson problem in 7.5 min for a mesh with 1.5 billion mesh cells. Standard methods (mldivide, operation in Matlab) and PCG (conjugate gradient

Table 1 Performance metrics for the pure diffusion problem. The relative residual is defined at $\| \frac{Lu-v}{v} \|$. "mldivide" is the standard matrix equation solver in Matlab, and PCG is the preconditioned conjugate gradient method. The preconditioning method used in PCG is the modified incomplete Cholesky factorization with no fill. An asterisk indicates that the solver did not converge in a reasonable time

Num voxels (millions)	Solution time (seconds)			Relative residual		
	Fast	mldivide	PCG	Fast	mldivide	PCG
0.01	**0.01**	**0.06**	**0.08**	1e–14	2e–14	1e–10
0.1	**0.02**	**1.4**	**0.3**	5e–14	7e–14	8e–11
1	**0.07**	**84**	**4**	2e–13	3e–13	8e–11
8.3	**0.4**	*	48	9e–13	*	9e–11
66.6	**3**	*	*	4e–12	*	*
535	**24**	*	*	2e–11	*	*
1002	**126**	*	*	4e–11	*	*
1499	**452**	*	*	4e–11	*	*

with modified incomplete Cholesky preconditioning) fail for domain sizes larger than 1 million equations.

We note in Table 1 that while the rate at which solution time increases for the fast solver is roughly linear for the first six rows, it increases sharply after transitioning past one billion voxels. This behavior can also be seen in the solution times for the largest network in Table 3 because the preconditioner also uses the discrete sine transforms that make up the fast solver. This increase in solution time is a result of the side lengths no longer being of the form $2^k - 1$. See Appendix 2 for a discussion and demonstration of this phenomenon.

3.2 *Idealized Exchange of Oxygen Across the Blood Brain Barrier (Linear Mass Transfer)*

We solve the preconditioned Schur complement system (31) associated with the networks in Fig. 3 in the case where hemoglobin bound O_2 is free to cross into extravascular space, and oxygen consumption by metabolism in the tissue is modelled by first-order reactions that is, the funcion F_v is set to be the identity function and $F_t(x) = kx$ for a constant k. The corresponding master equation (15) is thus linear.

We use GMRES(10) [i.e. GMRES with restart every 10 iterations] for varying resolutions (values of n_t and h) and the parameter values listed in Table 2.

The highest resolution we were able to achieve was just over one billion tissue voxels because of limited memory in our standard desk PCs (Intel Xeon W-2125 processor at 4.01 GHz and 256 GB of memory). We stop the iterative solver when the relative residual drops below 1e–6. In all cases the relative residual of the whole system was orders of magnitude smaller than this, so presumably many fewer

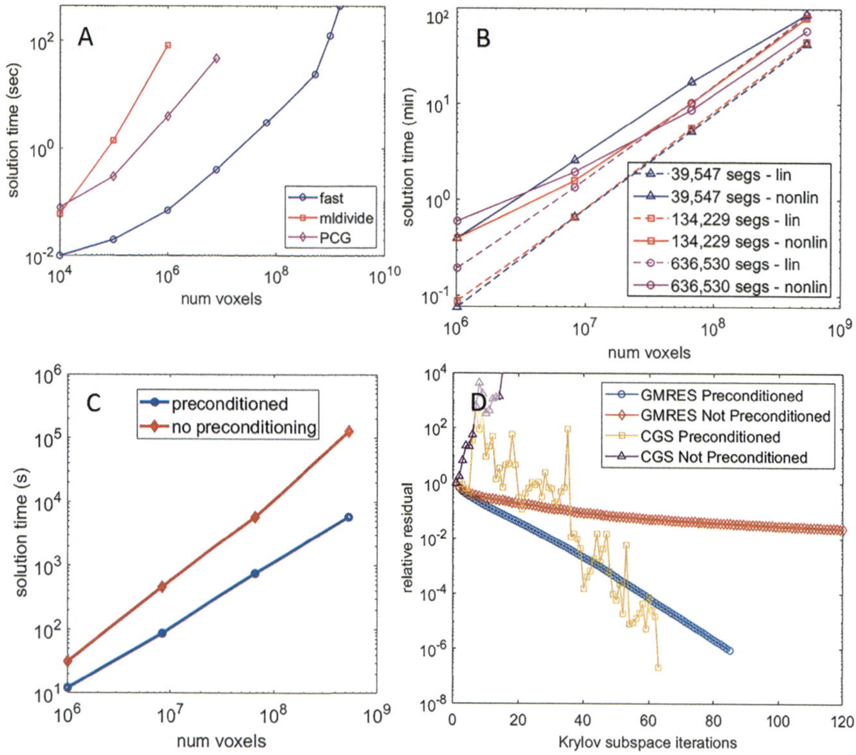

Fig. 4 (**a**) Comparison of solution times of the fast Poisson solver to a standard direct solver "mldivide" in Matlab and the iterative solver PCG (conjugate gradient with modified incomplete Cholesky preconditioning) for symmetric matrix systems. (**b**) Comparison of solution times between the linear and nonlinear formulations of the O2 exchange for the three networks in (**c**) Comparison of solution times for the linear problem with and without fast-Poisson preconditioning. (**d**) The fast-Poisson preconditioning has a nontrivial effect on the size of the relative residual of the linear problem at each iteration, for two different Krylov subspace methods for square non-symmetric systems: GMRES (generalized minimal residual) and CGS (conjugate gradient squared)

iterations are needed in practice. The time to solve for c_v once the solution c_t was in hand was negligible by comparison (less than one minute). Performance metrics for identical resolutions as those in Sect. 3.1 are reported in Table 3. A comparison of solution times is displayed in Fig. 4. We observe that faster convergence times are possible with a different Krylov subspace method; for instance, conjugate gradient squared (CGS) but decided to stick with GMRES because it was the most reliable, in contrast to CGS and other solvers which do not necessarily reduce the residual at each iteration.

Table 2 Parameter values used in simulation. Reaction rate was tweaked to obtain a minimum pO_2 of no less than 10 mmHg in each network, in agreement with the range in [17]

Parameter	Value	Units	Description	Reference
$\overline{\mathbf{p}}$	120, 5	mmHg	Inlet and outlet blood pressure respectively	[17]
C_0	0.02	mol/L	Bound O_2 concentration of fully saturated RBCs	[13]
n	2.59		Hill coefficient	[13]
P_{50}	40.2	mmHg	Dissolved O_2 partial pressure at 50% saturation	[13]
α	1.2e–6	mol/L/mmHg	Solubility of O_2 in plasma	[13]
$\overline{\mathbf{c}}_v$	68	mmHg	Arterial pO_2	[17]
$\overline{\mathbf{c}}_t$	40	mmHg	Ambient tissue pO_2 (constant Dirichlet boundary condition)	[15]
Γ_v	1.77e+03	$\mu m^2/s$	Permeability of vessel wall to O_2	[17]
h_v	1	μm	Vessel wall thickness	[17]
Γ_t	1.8e+03	$\mu m^2/s$	Diffusivity of O_2 in tissue	[25]
k	0.4, 0.07, 0.25	$1/s$	First order reaction rate (for networks A, B, and C in Fig. 1 respectively)	

3.3 Cortical Oxygen Extraction with Nonlinear Dissociation Kinetics

In this section we use three networks derived from synthetic and in-vivo data whose voxelized representations are shown in Fig. 3. The solution c_v is reflected in the coloration of the networks on the bottom row of this figure.

We apply Newton's method to the full nonlinear system with resolutions identical to those in Sects. 3.1 and 3.2 and use preconditioned GMRES(10) at each Newton step as described in Sect. 2.6.1. For the lowest resolution cases we use a constant vector for an initial guess; otherwise, we interpolate the solution from the closest lower resolution level to the higher resolution grid. For instance, the $n_t = 8.3$ million case uses the interpolated output of the $n_t = 1$ million case as its initial guess. Performance metrics are reported in Table 3. Solution time is cumulative, reflecting the time needed to compute the initial guess by solving the lower resolution cases. Planar cuts of the 3D solution in the representative $n_t = 66.6$ million case are shown in Figs. 5 and 6. Figure 5 shows that features of the tissue oxygen field are maintained between resolution levels, demonstrating mesh independence, and Fig. 6 shows the level of detail achieved at even modest solution times, as the nonlinear 66.6 million voxel case converges in only 9 min.

Physiological validation studies of our tissue O_2 model against two-photon imaging of the rat brain are presented in [17].

Table 3 Performance metrics for the preconditioned Schur system (31) in the linear case and Newton's method in the nonlinear case. Three different networks were placed in rectangular tissue domains with an equal number of voxels. For $k = 7, 8, 9, 10$, the number of voxels parallel to the x and y axes are $2^k - 1$, and the number of voxels parallel to the z axis is $2^{(k-1)} - 1$ since the performance of the FFTW algorithm is optimized for side lengths of this form [11]. The linear Schur complement system and Newton sub-problems were solved with GMRES(10) and stopped when the relative residual dropped below 1e-6. For the 1 billion voxel case we used GMRES(5) to circumvent memory constraints. Note the nonlinear solution time is shorter than the linear solution time for the largest network because fewer Newton iterations were required than for the other networks. MT (Mass Transfer) voxels refers to those voxels which intersect the surface of the vascular domain and are supplied with oxygen by mass transfer between domains

Network size (mm^3)	Num segments	Num voxels n_t (millions)	Num MT voxels (millions)	Voxel side length h (μm)	Newton iterations	Solution time—linear (min)	Solution time—Nonlinear (min)
$3 \times 3 \times 1.3$	636350	1	0.3	25	8	0.2	0.6
		8.3	1.2	13	3	1.4	2
		66.6	4.9	6.3	2	10.5	8.8
		535	21.5	3.2	2	85.6	58.7
		1002	**33.9**	**2**	**4**	**442 (7.4 hr)**	**435 (7.3 hr)**
$1.1 \times 1 \times 1.1$	**134229**	**1**	**0.2**	**19**	**8**	**0.09**	**0.4**
		8.3	**0.6**	**9.3**	**3**	**0.7**	**1.6**
		66.6	**2.5**	**4.7**	**3**	**5.6**	**10.4**
		535	**10.7**	**2.3**	**3**	**45**	**80.5**
$1.1 \times 1.5 \times 1.8$	39547	1	0.1	31	8	0.08	0.4
		8.3	0.4	15	6	0.7	2.6
		66.6	1.4	7.7	5	5.2	17.2
		535	5.7	3.8	3	43	87.2

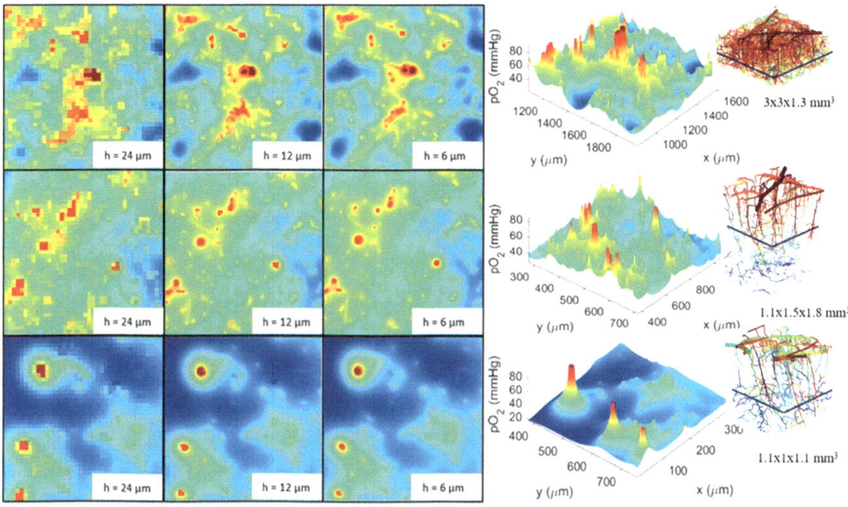

Fig. 5 Comparison of features of the tissue pO_2 field in a constant z slice at three different resolutions for three different networks. Features persist and become more sharply defined at higher resolutions, demonstrating mesh independence of the model. It is easy to see the characteristically steep and radially symmetric profile curves around penetrating arterioles when tissue pO_2 is plotted as a surface

4 Discussion

It is worth comparing the relationship between dual mesh techniques and 1d-3d coupling methods for transport developed by Zunino and others.

There are certain similarities between the two methods. Both techniques can be used to describe solute exchange between a thin tubular (vascular, 1D domain) structure and a surrounding continuum (3d tissue domain) leading to a set of linear algebraic systems of the PDE transport equations after discretization. They do not require a fine-grained contiguous mesh at the interface to resolve the neighborhood of the thin 1D domain, which would be prohibitive in cerebrovascular applications. The coarse resolution keeps the number of mesh cells within manageable limits. This is a critical advantage over Green's function methods [32] which involve a prohibitively large number of function evaluation for the 3d continuum, the Green's function method is essentially intractable for 3d computations. Moreover, both techniques rigorously enforce flux balances across the interface, which is necessary for conservation principles. This is accomplished naturally by enforcing flux equality over surfaces in the case of the dual mesh techniques. Because transfer fluxes can be defined over accurately resolved surfaces or surface projections beyond the resolution of the coarse grid interface, flux computations can be computed correctly even in meshes with coarse cells, or cells that fully blanket

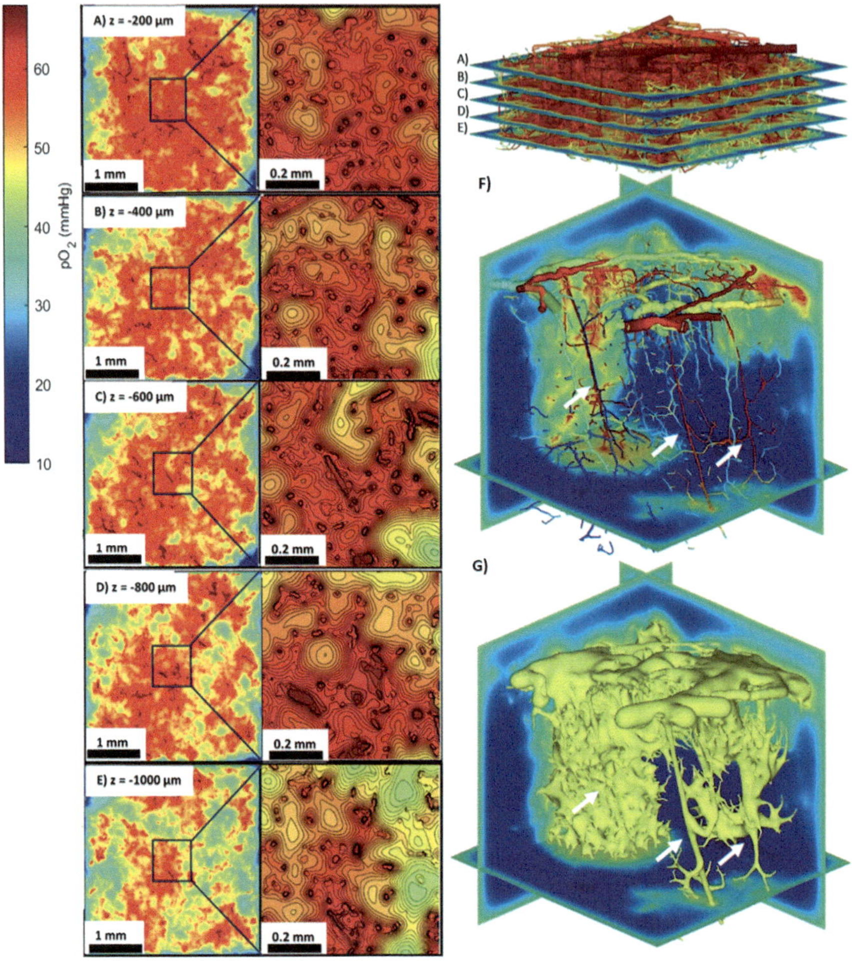

Fig. 6 (**a**)–(**e**) Horizontal cross-sections of O_2 field about the $3\times3\times1.3$ mm^3 synthetic network with voxel side length of 6 µm. Zoom in on features demonstrates the level of detail acquired at modest resolution and solution times (about 10 min) for large networks. (**f**) 3D rendering of the $1.1\times1\times1.1$ mm^3 mouse cortical network with vessels less than 5 µm in diameter omitted and cross-sections of corresponding O_2 field at 6 µm resolution. Penetrating arterioles are indicated with arrows. (**g**) The same view as in F with the 45 mmHg isosurface of the O_2 field displayed. The field clearly conforms to the shape of the network in 3D, especially near penetrating arterioles

thin 1D segments. This flux equality is equivalently enforced via the Lagrangian approach adopted in 1d-3d coupling.

However, significant differences exist between the two methods regarding the physics they describe and mathematical techniques to solve the underlying transport problems. 1d-3d coupling (e.g [21]) has been proposed to predict water exchange across the blood brain barrier driven solely by hydrostatic driving forces. However, water transport in the brain is intricately coupled to driving forces stemming from colloidal and osmotic pressure gradients across the BBB. Extravasation of water purely based on hydrostatic pressure gradients neglects this significant phenomenon of osmotic coupling. A physiologically sound hydrostatic/osmotic coupling has been described in Linninger et al. [23].

Under the problematic assumption of purely hydrostatic pressure driven transport, both the vascular and the tissue transport conform to diffusive transport leading to an elliptic set of transport equations in the tubular network as well as for the tissue. Hence the steady case, 1d-3d approach describes elliptic/elliptic coupling with no reactions in tissue. The 1d-3d method has typically applied to unstructured tetrahedral (finite element meshes) which leads to a system matrix with large bandwidth, a property know to be detrimental to the solvability of the resulting linear set of equations [28].

The dual mesh approach has been applied for the convective transport of solute in blood, such as oxygen, CO_2 or glucose, which cross the BBB driven by suitable concentration gradients with chemical reactions in tissue. Complex nonlinear dissociation kinetics in biphasic (plasma and RBC) flow has been incorporated [16, 34] for blood and non-linear Michaelis Menton kinetics for oxygen consumption in tissue [16, 34]. Thus, dual mesh applications have successfully dealt with convective solute transport in blood coupled with diffusive reactive phenomena in tissue. The resulting system of equations after discretization are hyperbolic for transport in blood coupled with elliptic equations with source terms for tissue. The Schur complement approach cleverly embeds the hyperbolic contribution of the vascular domain as a source term into the reactive tissue equations. Solute mass transfer from blood enters as a source term in the tissue domain. The implicit integration of convective source terms via the Schur compliment has not been incorporated in 1d-3d coupling, convective transport in blood has not been addressed in 1d-3d methods.

The dual mesh technique seamlessly extends to the dynamics case, in which it solves hyperbolic/elliptic coupling. Extensions to dynamic mass exchange are not reported for 1d-3d coupling, but have been demonstrated for impulse transfer from 3d to 1d networks models (see [31]). Moreover, the dual mesh techniques have been extended beyond unstructured tetrahedral meshes to apply to structured Cartesian space representations. In effect within Cartesian reference frames (from tissue with embedded thin tubular network), no mesh is needed at all, so that the method is also termed as mesh free. In fact, the solute problems can be solved using simple graph operators without the need for complex FEM code or associated unstructured meshes. Moreover, it should be noted that the use of tetrahedral meshes required in 1d-3d coupling is quite inefficient, since the purpose of coarse technique is to avoid the need for a (unstructured, body-fitted, contiguous interface) mesh to conform to the articulated vascular network structure. With tetrahedral meshes, this benefit is given away. For coarse domain representations, Cartesian tissue discretization is the most natural and computational most efficient implementation.

This has several advantages: the number of mesh cells is much smaller, the bandwidth is minimal and the resulting transport problem conforms to a canonical

Poisson problem, whose system matrix has known eigenvalue/eigenvectors. We have in this paper exploited this property to implement fast Fourier Poisson preconditioners that enable the solution of massive high resolution oxygen extraction problems with extreme speed. In contrast 1d/3d coupling has not been shown for more than a few hundred thousand equations.

For oxygen transport in the brain, the dual mesh technique has been demonstrated as method of choice when modeling large tissue domains. An interesting extension for simulation of transport phenomena for the whole brain would entail harnessing further computational gains by combining neuronal networks with ultrahigh resolution dual mesh techniques. This combined technique might enable whole brain simulation of oxygen extraction at high resolution.

5 Conclusions

High resolution (up to the micron level) simulation of cortical oxygen perfusion is necessary for physiologically meaningful analysis of the relationship of microvascular topology to neurovascular dysfunction. We presented a novel mesh-free mixed-domain technique based on fast discrete sine transfer (DST) preconditioning which enabled the modeling of oxygen perfusion in cerebrovascular systems of unprecedented high resolution and anatomical complexity. This allows us to resolve the smallest scales of the microvasculature in realistic networks down to the single micron scale. Large scale domain sizes are also essential for eliminating non-physical boundary effects; this can be achieved by simulating oxygen exchange over a larger region but restricting analysis only to a small core region of interest far removed from the domain boundaries.

Our innovations rested on a dual mesh technique which uses graph representations of the vascular and tissue domains and their coupling for automatic generation of model equations. A Cartesian reference mask for the tissue domain with embedded vascular network enables a simple coupling scheme that can resolve the smallest capillaries without the need for fitted mesh generation. Furthermore, the Cartesian structured matrices allowed us to leverage fast Fourier transform algorithms to solve closely related Poisson problems extremely quickly at ultra-high resolution, affording a cheap preconditioner for the tissue oxygen computation. Finally, the Schur complement implicitly incorporates the effect of mass transfer from vasculature into tissue and generates a fully coupled system that can be effectively preconditioned by the fast Poisson solver. Nonlinear dissociation and metabolic kinetics did not diminish the efficacy of the method, requiring only the additional complication of utilizing Newton's method in which the linear subproblem is solved at each iteration.

The preconditioned Schur complement method also works with very little modification in the dynamic case, enabling the high resolution study of oxygen fields over time. These encouraging results further motivate observing time dependent oxygen perfusion subject to stimuli such as neuronal firing or periodic excitation (e.g. vasomotor drive).

Appendix 1: Discretization of the Continuous Problem

For oxygen concentration c, the governing continuous equation is a generic convection-diffusion-reaction equation describing mass balance per unit volume per unit time at each point:

$$\frac{\partial c}{\partial t} = -\nabla \cdot (\mathbf{u}c) + \nabla \cdot (D\nabla c) - R(c) \tag{39}$$

where $\mathbf{u}$ is the velocity field, D is diffusivity, and $R(c) > 0$ is a metabolic reaction term.

Our overall domain Ω is a rectangle in 3D which is split into two closed regions: the vascular domain Ω_v is a 3D tubular network of blood vessels which is embedded in the tissue domain $\Omega_t = \Omega \setminus \text{int}(\Omega_v)$. The restrictions of c to the two domains are denoted respectively by $c_v = c \mid_{\Omega_v}$ and $c_t = c \mid_{\Omega_t}$.

We prescribe that c_v denotes the concentration of oxygen bound to hemoglobin, while c_t is the concentration of free oxygen in tissue. Consequently, there is a discontinuity for the value of c at the boundary $\partial\Omega_v$, although we preserve mass balance by prescribing that the mass flux between domains at the boundary satisfies

$$\nabla c \cdot \mathbf{n} = c_t - F_v(c_v), \qquad \text{on } \partial\Omega_v \tag{40}$$

where $\mathbf{n}$ is the outward pointing unit vector on the surface $\partial\Omega_v$, and F_v is the is the nonlinear function derived from Hill's equation which converts bound to free oxygen, given by Eq. (5).

To further distinguish between the domains, we prescribe that blood only flows in the vascular network, reactions only occur in the tissue, and diffusion is only present in the tissue and at the boundary between domains:

$$\mathbf{u} = 0 \text{ on the interior of } \Omega_t \tag{41}$$

$$R = \begin{cases} F_t(c_t) & \text{on } \Omega_t \\ 0 & \text{on the interior of } \Omega_v \end{cases} \tag{42}$$

$$D = \begin{cases} \Gamma_t & \text{on the interior of } \Omega_t \\ \Gamma_v & \text{on } \partial\Omega_v \\ 0 & \text{on the interior of } \Omega_v \end{cases} \tag{43}$$

where F_t is the Michaelis-Menten metabolic rate given by Eq. (14).

For boundary conditions, we prescribe a constant value for c_t at the boundary of Ω_t and for c_v at the vascular inlets.

Equations (39)–(43) for unknown c throughout the rectangle Ω together with the Dirichlet boundary conditions constitute the continuous problem. Our problem differs from the ones studied in, for instance [20] in the sense that oxygen in the vascular domain is transported by the bulk flow of blood; a hyperbolic phenomenon, in contrast to the elliptic Darcy flow which is absent in this domain. Furthermore, the continuous problem differs from others in the 1d-3d literature by being fully 3D: we do not seek to approximate a continuous function which couples PDEs across manifolds of different dimension.

Rather than "1d-3d coupling", a better descriptor of our approach is that of a *dual-mesh technique* (see [15, 17, 22]) where the choice of discretization differs between the two domains, introducing two distinct coupled meshes: a 1D graph for Ω_v and a traditional 3D Cartesian mesh of cell centered finite volumes for Ω_t. This choice of discretization for Ω_v will be explained presently.

The centerlines of the vascular domain Ω_v are given in the form of a spline network, and each spline is discretized into a sequence of straight line segments. In this way the skeleton of Ω_v is realized as a graph of nodes and edges, where every edge has a prescribed diameter and length. Each edge is associated with a capsule: a right-circular cylinder of the same diameter and length with hemispherical endcaps. A control volume is associated with each node of the graph by taking the union of the interiors of halves of all capsules converging at the node: see Fig. 1.

Integrating (39) over such a control volume yields by the divergence theorem a surface integral over the control volume surface:

$$V_{RBC}\frac{\partial c_v}{\partial t} = \oiint (D\nabla c_v - \mathbf{u}c_v)\cdot \mathbf{n}\, dS. \tag{44}$$

Since c_v is the concentration of oxygen bound to hemoglobin, we are only concerned with the fraction of the volume occupied by red blood cells: $V_{RBC} = HV$; where H is the hematocrit at the node and V is the geometric volume of the control volume.

Decompose the surface S of the control volume into $S = S_1 \cup \cdots \cup S_n \cup S_{MT}$ where S_i is the circular boundary with an adjacent control volume (and n is the degree of the node) and S_{MT} is the remainder of the surface: the boundary with Ω_t through which oxygen mass is transferred by diffusion. Since $D = 0$ on S_i and $D = \Gamma_v$ on S_{MT}, the Eq. (44) can be rewritten into:

$$V_{RBC}\frac{\partial c_v}{\partial t} = \oiint_{S_{MT}} (\Gamma_v\nabla c_v - \mathbf{u}c_v)\cdot \mathbf{n}\, dS - \sum_{i=1}^{n} \oiint_{S_i} (\mathbf{u}c_v)\cdot \mathbf{n}\, dS. \tag{45}$$

We assume that the flow field $\mathbf{u}$ at S_i is parallel to the graph edge and constant over S_i. Hence, if S_i is an inflow surface, then $\mathbf{u}$ is anti-parallel to $\mathbf{n}$ and id S_i is an outflow surface then $\mathbf{u}$ is parallel to $\mathbf{n}$. Similarly, $\mathbf{u}$ is perpendicular to $\mathbf{n}$ on S_{MT}.

Interpreting $\mathbf{u}$ as the velocity of RBCs, it follows that

$$\oiint_{\sigma}(-\mathbf{u}c_v)\cdot\mathbf{n}\,dS = \begin{cases} H_{\text{in}} f_{\text{in}} c_{v,\text{in}} & \text{if } \sigma = S_i \text{ is an inflow surface} \\ -H_{\text{out}} f_{\text{out}} c_{v,\text{out}} & \text{if } \sigma = S_i \text{ is an outflow surface} \\ 0 & \text{if } \sigma = S_{MT} \end{cases} \tag{46}$$

where f is the volumetric flow rate in the given edge and H is the hematocrit, so Hf is the net flow of RBCs in the edge. Following the upwinding scheme [26], we assign $c_{v,\text{in}}$ the average oxygen concentration in the upstream control volume, and $c_{v,\text{out}} = c_v$ the average concentration in the given control volume.

We further decompose S_{MT} into N pieces of equal area, where N is the number of tissue cells intersected by S_{MT}. We use a finite difference scheme to discretize the directional derivative $\nabla c_v \cdot \mathbf{n}$ over the surface. Hence, by the constraint (40), the mass transfer term becomes

$$\oiint_{S_{MT}} \Gamma_v \nabla c_n \cdot \mathbf{n}\,dS = \frac{\Gamma_v A_N}{h_v} \sum_{j=1}^{N} (c_{t,j} - F_v(c_v)) \tag{47}$$

where A_N is the area of the subsurface piece, h_v is the thickness of the vessel wall, and $c_{t,j}$ is the average concentration in an adjacent cell.

Equation (45) together with the discretizations (46) and (47) constitute the spatial discretization of the continuous equation on Ω_v; which in matrix form are the Eqs. (1) through (4). The discretization of Ω_t is a standard finite-volume scheme with cell-centered Cartesian cubic volume elements. We make the technical distinction that cells which intersect the boundary $\partial\Omega_v$ (IE, the surfaces of the cylindrical capsules) are supplied by the same mass-transfer term (47) as well as by diffusion from their neighbors and depletion from reactions. Cells which lie in the interior of Ω_v are overwritten with the values of their respective nodal control volumes [17].

Appendix 2: Dependence of the DST on Domain Side Length

Most of the work done in each iteration of GMRES in the solution of the preconditioned Schur complement system (31) is by the 3D discrete sine transforms in the preconditioner. To compute the discrete sine transform we use the open-source FFTW library [11]. The FFTW code uses a number of solvers and highly optimized FFT algorithms dependent on your specific system architecture. In general, if the input vector to a 1D DST has length n, then the 1D DST is most efficient if $N = 2(n + 1)$ is a product of small prime factors. Since a 3D DST takes successive 1D DSTs in the x, y, and z directions, we chose to optimize the DST by choosing $n = 2^k - 1$ for the number of cells in the x, y, and z directions so that N is a

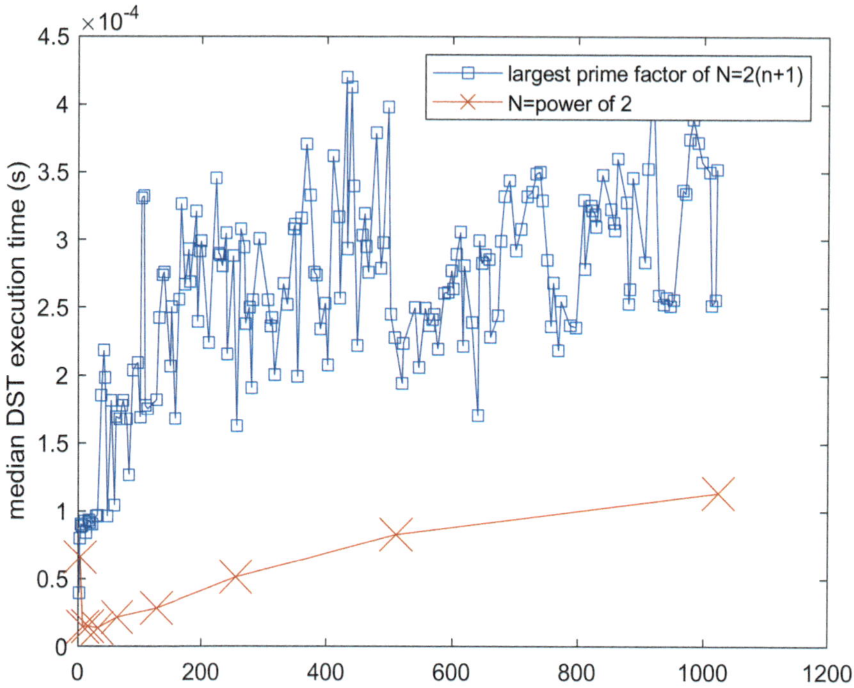

Fig. 7 Comparison of execution times for the 1D discrete sine transform. Xs are execution times for N = a power of 2. Ordinates of squares are median execution times for all input lengths for which the largest prime factor of N is the abscissa

power of two. We did this for all cases except those larger than one billion voxels, in which case we attempted to just get the highest resolution possible with the limited memory in our computer.

Because of this choice, we had less control over the size of the prime factors in the x, y, and z directions for the $>$ 1 billion voxels tests, and this increased the execution time in greater proportion than the trend observed in the powers of two. See Fig. 7 for a direct comparison of execution times for the DST when N is a power of two versus a product of large primes.

Appendix 3: Dynamic Oxygen Simulation for a Single Capillary

As a proof of concept for the dynamic simulation, we solve the fully nonlinear dynamic equations for a single capillary (single vascular segment) embedded in a tissue domain of 1 million voxels.

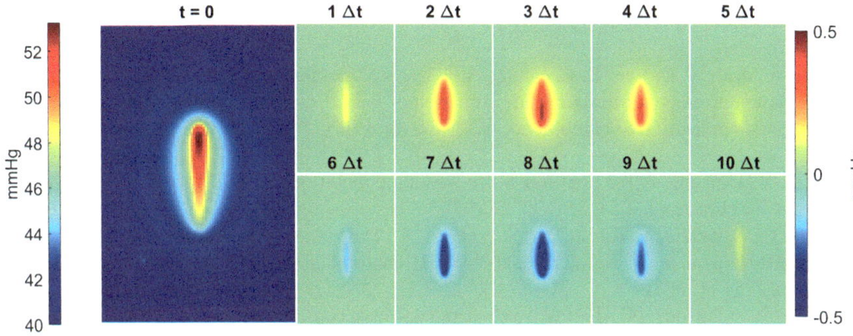

Fig. 8 Left: constant x-coordinate slice of steady-state tissue O_2 field c_t for single capillary. Right: difference between successive time-steps over a single cardiac cycle. One cardiac cycle is 10 time-steps

The inlet pressure was made to vary between 75 and 100 mmHg in a sinusoidal pattern with a frequency of 500 beats per minute: this was chosen to match observations of the heart-rate and blood pressure in resting mice [19]. For each distinct value of pressure the corresponding flow and hematocrit fields were computed using the method described in Sect. 2.3. We found that the hematocrit in each edge remained constant and the flow in each vascular edge varied in a sinusoidal pattern with the same frequency, so it was sufficient to precompute hematocrit and the minimum and maximum flow rates for each edge and then simply model the flow at each time step with a cosine wave. This saves time in the dynamic simulation since we do not need to recompute the flow and hematocrit at each time step.

The steady-state solution was computed then used as an initial condition for the dynamic simulation. See Fig. 8 for a display of the steady-state solution and the change in tissue oxygen tension between successive time steps over a single cardiac cycle. The tissue oxygen field follows the sinusoidal pattern of blood flow.

References

1. Axelsson, O.: Iterative Solution Methods. Cambridge University Press, Cambridge (1994). https://doi.org/10.1017/CBO9780511624100
2. Balay, S., Abhyankar, S., Adams, M., Brown, J., Brune, P., Buschelman, K., Dalcin, L., Dener, A., Eijkhout, V., Gropp, W., et al.: PETSc users manual (2019)
3. Bang-Jensen, J., Gutin, G.Z.: Digraphs: Theory, Algorithms and Applications. Springer, Berlin (2008)
4. Bertsekas, D.P.: Nonlinear programming. J. Oper. Res. Soc. **48**(3), 334–334 (1997)
5. Blinder, P., Tsai, P.S., Kaufhold, J.P., Knutsen, P.M., Suhl, H., Kleinfeld, D.: The cortical angiome: an interconnected vascular network with noncolumnar patterns of blood flow. Nat. Neurosci. **16**(7), 889–897 (2013)
6. Briggs, W.L., Henson, V.E., McCormick, S.F.: A Multigrid Tutorial. SIAM, Philadelphia (2000)

7. Butcher, J.C.: Numerical Methods for Ordinary Differential Equations. Wiley, New York (2016)
8. Celaya-Alcala, J.T., Lee, G.V., Smith, A.F., Li, B., Sakadžić, S., Boas, D.A., Secomb, T.W.: Simulation of oxygen transport and estimation of tissue perfusion in extensive microvascular networks: application to cerebral cortex. J. Cerebral Blood Flow Metabolism **41**(3), 656–669 (2021)
9. Chan, R.H., Wong, C.K.: Sine transform based preconditioners for elliptic problems. Numer. Linear Algebra Appl. **4**(5), 351–368 (1997)
10. Fang, Q., Sakadžić, S., Ruvinskaya, L., Devor, A., Dale, A.M., Boas, D.A.: Oxygen advection and diffusion in a three-dimensional vascular anatomical network. Optics Express **16**(22), 17530–17541 (2008)
11. Frigo, M., Johnson, S.G.: FFTW user's manual. Massachusetts Institute of Technology (1999)
12. Frigo, M., Johnson, S.G.: The design and implementation of FFTW3. Proc. IEEE **93**(2), 216–231 (2005)
13. Gagnon, L., Sakadžić, S., Lesage, F., Musacchia, J.J., Lefebvre, J., Fang, Q., Yücel, M.A., Evans, K.C., Mandeville, E.T., Cohen-Adad, J., et al.: Quantifying the microvascular origin of bold-fMRI from first principles with two-photon microscopy and an oxygen-sensitive nanoprobe. J. Neurosci. **35**(8), 3663–3675 (2015)
14. Golub, G.H., Van Loan, C.F.: Matrix computations. JHU Press, Baltimore (2013)
15. Gould, I.G., Linninger, A.A.: Hematocrit distribution and tissue oxygenation in large microcirculatory networks. Microcirculation **22**(1), 1–18 (2015)
16. Gould, I.G., Tsai, P., Kleinfeld, D., Linninger, A.: The capillary bed offers the largest hemodynamic resistance to the cortical blood supply. J. Cerebral Blood Flow Metabolism **37**(1), 52–68 (2017)
17. Hartung, G., Badr, S., Moeini, M., Lesage, F., Kleinfeld, D., Alaraj, A., Linninger, A.: Voxelized simulation of cerebral oxygen perfusion elucidates hypoxia in aged mouse cortex. PLoS Comput. Biol. **17**(1), e1008584 (2021)
18. Hill, A.V.: The possible effects of the aggregation of the molecules of hemoglobin on its dissociation curves. J. Physiol. **40**, iv–vii (1910)
19. Ho, D., Zhao, X., Gao, S., Hong, C., Vatner, D.E., Vatner, S.F.: Heart rate and electrocardiography monitoring in mice. Curr. Protocols Mouse Biol. **1**(1), 123–139 (2011)
20. Koch, T., Schneider, M., Helmig, R., Jenny, P.: Modeling tissue perfusion in terms of 1d-3d embedded mixed-dimension coupled problems with distributed sources. J. Comput. Phys. **410**, 109370 (2020)
21. Köppl, T., Vidotto, E., Wohlmuth, B., Zunino, P.: Mathematical modeling, analysis and numerical approximation of second-order elliptic problems with inclusions. Math. Models Methods Appl. Sci. **28**(05), 953–978 (2018)
22. Linninger, A., Gould, I., Marinnan, T., Hsu, C.Y., Chojecki, M., Alaraj, A.: Cerebral microcirculation and oxygen tension in the human secondary cortex. Ann. Biomed. Eng. **41**, 2264–2284 (2013)
23. Linninger, A.A., Xu, C., Tangen, K., Hartung, G.: Starling forces drive intracranial water exchange during normal and pathological states. Croatian Med. J. **58**(6), 384–394 (2017)
24. Mandel, J.: On block diagonal and schur complement preconditioning. Numer. Math. **58**(1), 79–93 (1990)
25. Mintun, M.A., Lundstrom, B.N., Snyder, A.Z., Vlassenko, A.G., Shulman, G.L., Raichle, M.E.: Blood flow and oxygen delivery to human brain during functional activity: theoretical modeling and experimental data. Proc. Natl. Acad. Sci. **98**(12), 6859–6864 (2001)
26. Patankar, S.: Numerical Heat Transfer and Fluid Flow. Taylor & Francis, Milton Park (2018)
27. Quarteroni, A., Sacco, R., Saleri, F.: Numerical Mathematics, vol. 37. Springer, Berlin (2010)
28. Rosen, R.: Matrix bandwidth minimization. In: Proceedings of the 1968 23rd ACM National Conference, pp. 585–595 (1968)
29. Saad, Y.: Iterative Methods for Sparse Linear Systems. SIAM, Philadelphia (2003)
30. Saad, Y., Schultz, M.H.: GMRES: A generalized minimal residual algorithm for solving nonsymmetric linear systems. SIAM J. Sci. Stat. Comput. **7**(3), 856–869 (1986)

31. Salvador, M., Dede', L., Quarteroni, A.: An intergrid transfer operator using radial basis functions with application to cardiac electromechanics. Comput. Mech. **66**, 491–511 (2020)
32. Secomb, T.W., Hsu, R., Park, E.Y., Dewhirst, M.W.: Green's function methods for analysis of oxygen delivery to tissue by microvascular networks. Ann. Biomed. Eng. **32**, 1519–1529 (2004)
33. Treeby, B.: Matlab discrete trigonometric transform library (2020). https://github.com/ucl-bug/matlab-dtts/releases/tag/v1.1
34. Ventimiglia, T., Linninger, A.A.: Mesh-free high-resolution simulation of cerebrocortical oxygen supply with fast fourier preconditioning. Int. J. Numer. Methods Biomed. Eng. **39**, e3735 (2023)

Robust Preconditioning of Mixed-Dimensional PDEs on $3d$-$1d$ Domains Coupled with Lagrange Multipliers

Nunzio Dimola, Miroslav Kuchta, Kent-Andre Mardal, and Paolo Zunino

Abstract In the context of micro-circulation, the coexistence of two distinct length scales—the vascular radius and the tissue/organ scale—with a substantial difference in magnitude, poses significant challenges. To handle slender inclusions and simplify the geometry involved, a technique called *topological dimensionality reduction* is used, which suppresses the manifold dimensions associated with the smaller characteristic length. However, the algebraic structure of the resulting discretized system presents a challenge in constructing efficient solution algorithms. This chapter addresses this challenge by developing a robust preconditioner for the $3d$-$1d$ problem using the operator preconditioning technique. The robustness of the preconditioner is demonstrated with respect to the problem parameters, except for the vascular radius. The vascular radius, as demonstrated, plays a fundamental role in the mathematical well-posedness of the problem and the effectiveness of the preconditioner.

Keywords Mixed-dimensional equations · Saddle-point problems · Preconditioning

---

N. Dimola · P. Zunino (✉)
MOX, Department of Mathematics, Politecnico di Milano, Milan, Italy
e-mail: nunzio.dimola@polimi.it; paolo.zunino@polimi.it

M. Kuchta
Simula Research Laboratory, Oslo, Norway
e-mail: miroslav@simula.no

K.-A. Mardal
Numerical Analysis and Scientific Computing, Simula Research Laboratory, Oslo, Norway

Department of Mathematics, University of Oslo, Oslo, Norway
e-mail: kent-and@math.uio.no

A. Linninger et al. (eds.), *Quantitative Approaches to Microcirculation*,
SEMA SIMAI Springer Series 36, https://doi.org/10.1007/978-3-031-58519-7_6

1 Introduction

The human cardiovascular system displays a diverse range of scales and characteristics, encompassing the major blood vessels, arterioles, and capillaries, with diameters ranging from several cm to a few μm. In particular, when examining the microcirculation, the challenge intensifies due to the substantial disparity in scale between the diameter of small vessels (μm) and the size of the corresponding system or organ they supply (dm). To deal with this disparity, intricate vascular networks that occupy space are needed, leading to complex geometric structures.

Due to the inherent complexities involved, simulating flow throughout the cardiovascular system is not practical or applicable. However, it is crucial to adopt a comprehensive approach that incorporates interactions between different system components by integrating models operating at various levels of detail. This approach is commonly referred to as *geometrical multi-scale* or *hybrid-dimensional modeling*.

Two types of hybrid-dimensional models can be considered: sequential and embedded. In an embedded multiscale model, components of varying levels of detail are integrated within the same domain. A prime example is the microcirculation, where a complex vascular network comprising arterioles, capillaries, and small veins exists within the biological tissue.

From a modeling point of view, such ideas have appeared in the last three decades (at least) for modeling wells in subsurface reservoirs in [39, 40] and for modeling microcirculation in [3, 13, 14, 19, 45]. A similar approach has recently been used to model soil/root interactions [22]. However, these application-driven seminal ideas were not followed by a systematic theory and rigorous mathematical analysis. At the same time, the models introduce additional mathematical complexity, in particular concerning the functional setting for the solution as it involves coupling PDEs on domains with high dimensionality gap. This mathematical challenge has recently attracted the attention of many researchers. The sequence of works by D'Angelo [10, 11] and D'Angelo and Quarteroni [12], followed by Laurino and Zunino [30], Köppl et al. [25], and Kuchta et al. [29], has remedied the well-posedness by weakening the regularity assumptions that define a solution.

However, the mathematical understanding of mixed-dimensional embedded problems is not sufficient to successfully apply these models to realistic problems. An essential difficulty to overcome is the development of efficient numerical solvers that can handle the large separation of spatial scales between the domains of the equations and the subsequent geometrical complexity of the vascular network.

Actually, the study of the interplay between the mathematical structure of the problem and solvers, as well as preconditioners for its discretization, is still in its infancy. The results presented in [27] for the solution of the $1d$ differential equations embedded in $2d$, and more recently extended to the $3d$-$1d$ case in [28], have paved the way, but much more must be understood. This work proceeds in this direction, with the aim of discussing the main mathematical challenges at the basis of the

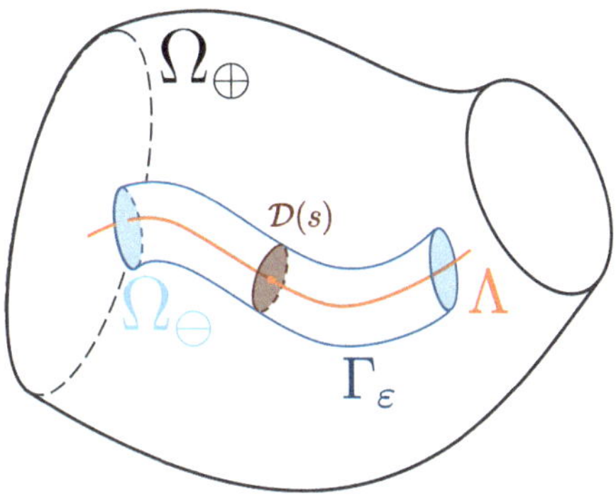

Fig. 1 Geometrical setting considered in this paper, and the associated notation. Domain $\Omega_\ominus$ is assumed to be a slender generalized cylinder characterized by the centerline Λ parameterized by the arclength coordinate s and $2d$ transversal cross section $\mathcal{D}(s)$

development of optimal solvers of mixed-dimensional $3d$-$1d$ problems coupled by Lagrange multipliers.

We model the case of one single small vessel by the domain $\Omega \subset \mathbb{R}^3$ that is an open, connected, and convex set that can be subdivided in two parts, $\Omega_\ominus$ and $\Omega_\oplus = \Omega \setminus \overline{\Omega}_\ominus$ representing respectively the vessel and the tissue. Let $\Omega_\ominus$ be a *generalized cylinder* that is the swept volume of a two-dimensional set, $\partial\mathcal{D}$, moved along a curve, Λ, in the three-dimensional domain, Ω, see Fig. 1 for an illustration. For dimensional reduction of the vessel domain we finally assume that $|\mathcal{D}| \ll |\Omega|$.

With the purpose of developing robust preconditioners for the *3d-1d* problem arising from microcirculation, let us consider a prototype problem that originates from coupling of two $3d$ second order elliptic equations:

$$
\begin{aligned}
-\kappa_\oplus \Delta u_\oplus &= f && \text{in } \Omega_\oplus,\\
-\kappa_\ominus \Delta u_\ominus &= f && \text{in } \Omega_\ominus,\\
u_\oplus - u_\ominus &= g && \text{on } \Gamma_\varepsilon,\\
\kappa_\oplus \nabla u_\oplus \cdot \nu_{\Gamma_\varepsilon} - \kappa_\ominus \nabla u_\ominus \cdot \nu_{\Gamma_\varepsilon} &= 0 && \text{on } \Gamma_\varepsilon,\\
u_\oplus,\ u_\ominus &= 0 && \text{on } \partial\Omega.
\end{aligned}
\tag{1}
$$

Here, $u_\oplus$ and $u_\ominus$ are the unknowns, and $\kappa_\oplus > 0$ and $\kappa_\ominus > 0$ are the diffusivities in the two different domains that we shall assume in the following to be constant for simplicity. Furthermore, $\Gamma_\varepsilon = \partial\Omega_\oplus \cap \partial\Omega_\ominus$ where $\varepsilon = \text{diam}\mathcal{D}$ denotes the diameter of the $2d$ transverse cross sections of $\Omega_\ominus$. Finally, f is a forcing term defined in whole Ω and ν_{Γ_ε} denotes the unit normal vector to the surface. The model presented in Eq. (1) is a simple template for the interaction of an interior and an exterior domain across the interface Γ_ε. In the context of microcirculation, it appears in many scenarios. For example, it can be used to describe the interaction between blood pressure in small vessels and interstitial fluid pressure. In this case, the variables $[u_\oplus, u_\ominus]$ represent the hydrostatic pressures of a fluid. It can also be adopted to describe the diffusion of chemical species with a small molecular weight (such as oxygen) from the vascular to the extravascular domains. In all cases, we point out that it represents a simplification of more complex phenomena. In the context of oxygen transport, which will be taken as a reference application throughout this work, it is well known that a diffusion-dominated model is not enough to capture the multiphysics nature of blood circulation; indeed, oxygen and hematocrit transport play a significant role in describing the phenomenology. Nevertheless, many numerical approaches for microcirculation tend to solve the equations for oxygen diffusion, convection, and hematocrit separately and in a segregated fashion, see, for example, [18]. In this prospective, the model presented has to be considered as a component of a more comprehensive model. Model (1) also deserves some consideration about the formulation of the coupling conditions. From a physical point of view, $u_\oplus - u_\ominus = g$ on Γ_ε imposes that the oxygen concentration (or partial pressure) is discontinuous throughout the vascular wall. These are the well-known Dirichlet coupling conditions that enforce the value of the solution on some defined portion of the boundary/interface. They are complemented by Neumann-type interface conditions that prescribe the continuity of normal fluxes, that is, $\kappa_\oplus \nabla u_\oplus \cdot \nu_{\Gamma_\varepsilon} - \kappa_\ominus \nabla u_\ominus \cdot \nu_{\Gamma_\varepsilon} = 0$. Dirichlet-Neumann interface conditions are the most fundamental coupling conditions of second-order elliptic PDEs. For this reason, they are our choice here. However, for transport across the wall of small vessels, the model most commonly studied is the Robin coupling [12], which describes the flux at the vessel interface between the vascular and extravascular domains $\Omega_\ominus$, $\Omega_\oplus$, respectively, by the following conditions defined at the interface Γ:

$$
\begin{aligned}
-\nabla u_\oplus \cdot \nu_{\Gamma_\varepsilon} &= \kappa (u_\oplus - u_\ominus), \\
-\nabla u_\ominus \cdot (-\nu_{\Gamma_\varepsilon}) &= \kappa (u_\ominus - u_\oplus).
\end{aligned}
$$

Normal oxygen flux at the interface is regulated by the parameter κ, which modulates the permeability of the vessel wall and the relative difference in oxygen concentration between the tissue and the interior of the vessel. However, as it is possible to interpret Robin conditions as a perturbation of the Dirichlet ones, our setting is also relevant for the Robin coupling; see [4].

Imposing Dirichlet boundary conditions via the Lagrange multiplier method can be thought of as imposing boundary conditions on Γ_ε weakly. With reference to Fig. 1, and considering the full-dimensional problem (1) for more clarity, the Dirichlet boundary terms coming from the weak formulation of the PDE are replaced by Lagrange multipliers $p_\oplus, p_\ominus \in H^{-1/2}(\Gamma_\varepsilon)$ such that

$$
\begin{aligned}
\int_{\Gamma_\varepsilon} p_\oplus \, \mathcal{T} v_\oplus \, d\sigma &:= \int_{\Gamma_\varepsilon} (\kappa_\oplus \nabla u_\oplus \cdot \nu_{\Gamma_\varepsilon}) \, \mathcal{T} v_\oplus \, d\sigma \quad \forall v_\oplus \in H_0^1(\Omega_\oplus), \\
\int_{\Gamma_\varepsilon} p_\ominus \, \mathcal{T} v_\ominus \, d\sigma &:= \int_{\Gamma_\varepsilon} (\kappa_\ominus \nabla u_\ominus \cdot \nu_{\Gamma_\varepsilon}) \, \mathcal{T} v_\ominus \, d\sigma \quad \forall v_\ominus \in H^1(\Omega_\ominus).
\end{aligned}
\tag{2}
$$

At this point, introducing the notation $(\cdot,\cdot)_\Sigma$ and $\langle\cdot,\cdot\rangle_\Sigma$ for the L^2 product and duality pairing in domain Σ, the variational formulation of (1) reads:

$$
\begin{aligned}
&(\kappa_\oplus \nabla u_\oplus, \nabla v_\oplus)_{\Omega_\oplus} + (\kappa_\ominus \nabla u_\ominus, \nabla v_\ominus)_{\Omega_\ominus} + \\
&+ \langle p_\oplus, \mathcal{T} v_\oplus\rangle_{\Gamma_\varepsilon} - \langle p_\ominus, \mathcal{T} v_\ominus\rangle_{\Gamma_\varepsilon} = (f, v_\oplus)_{\Omega_\oplus} + (f, v_\ominus)_{\Omega_\ominus}.
\end{aligned}
$$

By considering that on Γ_ε we have,

$$
\kappa_\oplus \nabla u_\oplus \cdot \nu_{\Gamma_\varepsilon} - \kappa_\ominus \nabla u_\ominus \cdot \nu_{\Gamma_\varepsilon} = 0 \iff \quad p_\oplus = p_\ominus := p,
$$

and introducing the weakly imposed Dirichlet boundary constraint

$$
\langle \mathcal{T}(u_\oplus - u_\ominus), q\rangle_{\Gamma_\varepsilon} = \langle g, q\rangle_{\Gamma_\varepsilon} \quad \forall q \in H^{-1/2}(\Gamma_\varepsilon), \tag{3}
$$

we obtain the following saddle-point problem

$$
\begin{aligned}
a([u_\oplus, u_\ominus], [v_\ominus, v_\ominus])_\Omega + b([v_\oplus, v_\ominus], p)_{\Gamma_\varepsilon} &= \langle f, [v_\oplus, v_\ominus]\rangle_\Omega \\
b([u_\oplus, u_\ominus], q)_{\Gamma_\varepsilon} &= \langle g, q_\odot\rangle_{\Gamma_\varepsilon}
\end{aligned} \tag{4}
$$

for all $[v_\oplus, v_\ominus] \in [H_0^1(\Omega_\oplus), H^1(\Omega_\ominus)]$, $q \in H^{-1/2}(\Gamma_\varepsilon)$, where

$$
\begin{aligned}
a([u_\oplus, u_\ominus], [v_\oplus, v_\ominus]) &= \kappa_\oplus(\nabla u_\oplus, \nabla u_\ominus)_{\Omega_\oplus} + \kappa_\ominus(\nabla u_\ominus, \nabla v_\ominus)_{\Omega_\ominus}, \\
b([v_\oplus, v_\ominus], q) &= \langle \mathcal{T}(v_\oplus - v_\ominus), q\rangle_{\Gamma_\varepsilon}.
\end{aligned}
$$

From the above analysis, we have recovered the saddle-point structure that is inherited by the reduced-dimensionality weak problem (6) from the full-dimensional one. In this context, the Lagrange multiplier p represents the weighted normal flux $\kappa \nabla u \cdot \nu_{\Gamma_\varepsilon}$ of the variable u. The formulation with the Lagrange multiplier as an auxiliary unknown that enforces the coupling is not the only option when dealing with Dirichlet conditions on the interface. The problem could be addressed, for example, by variants of the penalty method [35] or Nitsche's method [38]. We note that these methods introduce additional (stabilization) parameters into the variational problem.

By applying a suitable model reduction strategy to (1), which exploits the following transverse averages as already described in [30],

$$
\overline{w}(s) = |\partial \mathcal{D}(s)|^{-1} \int_{\partial \mathcal{D}(s)} w d\gamma \quad \text{and} \quad \overline{\overline{w}}(s) = |\mathcal{D}(s)|^{-1} \int_{\mathcal{D}(s)} w d\sigma,
$$

we obtain the *3d-1d* coupled problem of the form

$$\begin{aligned} -\kappa\Delta u + \varepsilon p_{\odot}\delta_{\Lambda} &= f && \text{in } \Omega, \\ -\varepsilon^2 \kappa_{\odot} d_s^2 u_{\odot} - \varepsilon p_{\odot} &= \varepsilon^2 \overline{\overline{f}} && \text{on } \Lambda, \\ \varepsilon(\overline{\mathcal{T}}_{\Lambda} u - u_{\odot}) &= g && \text{on } \Lambda, \\ u &= 0 && \text{on } \partial\Omega, \\ u_{\odot} &= 0 && \text{on } \partial\Lambda. \end{aligned} \tag{5}$$

Here $\kappa = \kappa_{\oplus}$ in $\Omega_{\oplus}$ and $\kappa_{\odot} = \overline{\overline{\kappa}}_{\ominus}$ in Λ being the centerline of $\Omega_{\ominus}$. Throughout the paper we will use the subscript $\odot$, e.g. $u_{\odot}$, for functions on Λ (and, in general, on domains with topological dimension smaller than that of the ambient space), while functions on Ω are denoted, as usual, by italic lowercase letters. Here, the primal unknowns are u and $u_{\odot}$, while the unknown $p_{\odot}$ is the Lagrange multiplier used to enforce the coupling of u and $u_{\odot}$. As a consequence of the dimensionality reduction technique, the Lagrange multiplier $p_{\odot}$ represents the cross-sectional (perimeter) average of the normal flux of u normal to the vessel interface. Furthermore, δ_{Λ} is a Dirac delta function. The operator obtained from a combination of the average operator $\overline{(\cdot)}$ with the trace in Γ_{ε} will be denoted by $\overline{\mathcal{T}}_{\Lambda} = \overline{(\cdot)} \circ \mathcal{T}_{\Gamma}$, since it maps functions in Ω to functions in Λ. Note that $\overline{\mathcal{T}}_{\Lambda}$ thus implicitly depends on ε.

Due to the high-dimensional gap between $3d$ and $1d$, and the presence of singularities in the solution u due to the $1d$ coupling, the well-posedness of the problem (5) is challenging [11, 12]. In the context of Robin coupling conditions at the $3d$-$1d$ interface, the singularity is addressed in a number of works, e.g. by proposing a splitting strategy where the singularity is captured by Green's functions [17], regularizing the problem by distributing the singular source term on the coupling surface or in the bulk [23, 25] or deriving error estimates in tailored norms [11, 24]. For these formulations, efficient solution algorithms have been developed utilizing algebraic multigrid [9, 21] or a fast Fourier transform for preconditioning [47].

Although the Robin coupling typically leads to elliptic problems, enforcing the Dirichlet coupling in (5) by Lagrange multiplier yields a saddle point system. Here, as recently shown in [29], the high $3d$-$1d$ dimensional gap requires the multiplier to reside in fractional order Sobolev spaces on Λ in order to obtain well-posedness. However, scalable solvers for the resulting linear systems are currently lacking. Here, our objective is to address this issue by constructing robust block-diagonal preconditioners for (5) within the framework of operator preconditioning [33]. In particular, to obtain robust estimates we establish the well-posedness of the coupled problem in weighted, fractional Hilbert spaces.

The structure of the chapter is described below. We start with the analysis of the coupled problem in Sect. 2. Next, we investigate the performance of the resulting preconditioner in Sect. 3, leading us to the observation that the parameter ε in (5) assumes a distinctive role beyond that of a standard material parameter. This unique role is further examined in Sect. 4, where we establish a close connection between ε and the well-posedness of the coupled problem. Finally, we provide our conclusions in Sect. 5.

2 Mathematical Formulation of the *3d-1d* Coupled Problem

Before delving into the precise mathematical formulation of problem (5), we will introduce some fundamental notation. Let D represent a bounded domain in $\mathbb{R}^d$, where $d = 1, 2, 3$. In this context, $L^2(D)$ denotes the space of functions that are square integrable in D, while $H^k(D)$ refers to the space of functions that possess k derivatives in L^2. We denote the closure of $C_0^\infty(D)$ in $H^k(D)$ as $H_0^k(D)$. When the context is clear, we may simplify the notation using H^k instead of $H^k(D)$. For a Hilbert space $X = X(D)$ with its corresponding dual space X', the norm and the associated inner products are denoted as $\|\cdot\|_X$ and $(\cdot, \cdot)_X$, respectively. The duality pairing is represented by $\langle \cdot, \cdot \rangle_X$, and if necessary, we will explicitly indicate the underlying domain using $(\cdot, \cdot)_D$ and $\langle \cdot, \cdot \rangle_D$. With slight abuse of notation, the dual of H^k is H^{-k} for $k \in \mathbb{R}$. Similarly, H_0^{-k} is the dual of H_0^k. We will use scaled and intersected inner products, norms, and Hilbert spaces for our analysis of robustness. For example $\varepsilon H_0^1(D)$ is defined in terms of the inner product

$$(u, v)_{\varepsilon H_0^1(D)} = \varepsilon^2 (\nabla u, \nabla v)_{L^2(D)}.$$

Notice that with this definition, the norm scales linearly in ε, that is, $\|u\|_{\varepsilon H1_0} = \varepsilon \|u\|_{H_0^1}$.

The intersection of two Hilbert spaces X and Y is denoted $X \cap Y$ and has an inner product

$$(x, y)_{X \cap Y} = (x, y)_X + (x, y)_Y.$$

We can now precisely formulate the weak formulation of the problem (5) that will be considered in this paper. The problem reads as follows: Find $u \in V$, $u_\odot \in V_\odot$, $p_\odot \in Q_\odot$ such that

$$\begin{aligned} a([u, u_\odot], [v, v_\odot]) + b([v, v_\odot], p_\odot) &= \langle f, [v, v_\odot] \rangle_\Omega & \forall [v, v_\odot] \in [V, V_\odot] \\ b([u, u_\odot], q_\odot) &= \langle g, q_\odot \rangle_\Lambda & \forall q_\odot \in Q_\odot. \end{aligned} \tag{6}$$

with

$$\begin{aligned} a([u, u_\odot], [v, v_\odot]) &= \kappa (\nabla u, \nabla u_\odot)_{L^2(\Omega)} + \kappa_\odot \varepsilon^2 (d_s u_\odot, d_s v_\odot)_{L^2(\Lambda)}, \\ b([v, v_\odot], q_\odot) &= \varepsilon \langle \overline{\mathcal{T}}_\Lambda v - v_\odot, q_\odot \rangle_\Lambda \end{aligned}$$

and $V = \kappa^{1/2}H_0^1(\Omega)$, $V_\odot = \kappa_\odot^{1/2}\varepsilon H_0^1(\Lambda)$, and $Q_\odot = \kappa^{-1/2}\varepsilon H^{-\frac{1}{2}}(\Lambda) \cap \kappa_\odot^{-1/2}H^{-1}(\Lambda)$. We remark that a main difference between [29] and the functional setting considered here is that the Lagrange multiplier space $Q_\odot$ is a weighted intersection space $\kappa^{-1/2}\varepsilon H^{-\frac{1}{2}}(\Lambda) \cap \kappa_\odot^{-1/2}H^{-1}(\Lambda)$ rather than $H^{-\frac{1}{2}}(\Lambda)$. This choice is fundamental for the development of robust preconditioners, as will be discussed later on.

2.1 The Stability of the Continuous Problem

We show well-posedness of the variational problem (6) in space $X = V \times V_\odot \times Q_\odot$ considered with its canonical norms by applying the abstract Brezzi theory [6]. To this end, we let

$$A : (V \times V_\odot) \to (V \times V_\odot)', \quad \langle A[u, u_\odot], [v, v_\odot]\rangle_{V\times V_\odot} = a([u, u_\odot], [v, v_\odot]),$$
$$B : (V \times V_\odot) \to Q_\odot', \quad \langle B[u, u_\odot], q_\odot\rangle_{Q_\odot} = b([u, u_\odot], q_\odot),$$

such that (6) can be equivalently stated as: Find $x = [u, u_\odot, p_\odot] \in X$ such that

$$\mathcal{A}x = f \text{ in } X', \quad \mathcal{A} = \begin{pmatrix} A & B' \\ B & 0 \end{pmatrix}.$$

The operator $\mathcal{A} : X \to X'$ is then an isomorphism (equivalently, (6) is well posed) provided that it satisfies the four Brezzi conditions, that is, the operators A and B are bounded, A is coercive on $\ker B \subset V \times V_\odot$ and B satisfies the inf-sup condition.

Of the four Brezzi conditions required for the problem (6), the coercivity and boundedness of A were established in [29] in the setting required here ($V \times V_\odot$ are the same and only $Q_\odot$ has changed) and will therefore not be repeated. Therefore, we only need to verify the properties of the B operator.

Theorem 1 $b(\cdot, \cdot) : [V, V_\odot] \times Q_\odot \to \mathbb{R}$ *satisfies the following conditions:*

$$b([u, u_\odot], q_\odot) \le C_B\|[u, u_\odot]\|_{[V,V_\odot]}\|q_\odot\|_{Q_\odot}, \quad u \in V, u_\odot \in V_\odot, q_\odot \in Q_\odot, \tag{7}$$

$$\sup_{[u,u_\odot]\in[V,V_\odot]} \frac{b([u, u_\odot], q_\odot)}{\|[u, u_\odot]\|_{[V,V_\odot]}} \ge \beta\|q_\odot\|_{Q_\odot}, \quad q_\odot \in Q_\odot \tag{8}$$

with positive constants C_B, β.

Proof We recall that owing to Dirichlet boundary conditions for u, $u_\odot$ we have the equivalence between the seminorms (induced by A) and the full H^1 norms. In [29] the following was established:

$$\varepsilon\|\overline{\mathcal{T}}_\Lambda u\|_{H^{1/2}(\Lambda)} \leq C_{\boldsymbol{T}}\|\boldsymbol{T}_{\Gamma_\varepsilon} u\|_{H^{1/2}(\Gamma_\varepsilon)} \leq C\|u\|_{H^1(\Omega)}, \quad \forall u \in H^1(\Omega),$$

where constants $C_{\boldsymbol{T}}$ and C are independent from ε. Hence, the boundedness of $b(\cdot,\cdot)$ can be obtained in a direct way as:

$$\begin{aligned}
b([u, u_\odot], q_\odot) &= \varepsilon\langle\overline{\mathcal{T}}_\Lambda u - u_\odot, q_\odot\rangle_\Lambda \\
&\leq \varepsilon|\langle\overline{\mathcal{T}}_\Lambda u, q_\odot\rangle_\Lambda| + \varepsilon|\langle u_\odot, q_\odot\rangle_\Lambda| \\
&\leq \|\overline{\mathcal{T}}_\Lambda u\|_{H^{1/2}(\Lambda)}\varepsilon\|q_\odot\|_{H^{-1/2}(\Lambda)} + \varepsilon\|u_\odot\|_{H^1(\Lambda)}\|q_\odot\|_{H^{-1}(\Lambda)} \\
&= \kappa^{1/2}\varepsilon\|\overline{\mathcal{T}}_\Lambda u\|_{H^{1/2}(\Lambda)}\kappa^{-1/2}\|q_\odot\|_{H^{-1/2}(\Lambda)} \\
&\quad + \varepsilon\kappa_\odot^{1/2}\|u_\odot\|_{H^1(\Lambda)}\kappa_\odot^{-1/2}\|q_\odot\|_{H^{-1}(\Lambda)} \\
&\leq \kappa^{1/2}C\|u\|_{H^1(\Omega)}\kappa^{-1/2}\|q_\odot\|_{H^{-1/2}(\Lambda)} \\
&\quad + \varepsilon\kappa_\odot^{1/2}\|u_\odot\|_{H^1(\Lambda)}\kappa_\odot^{-1/2}\|q_\odot\|_{H^{-1}(\Lambda)} \\
&\leq C_B(\|u\|_V^2 + \|u_\odot\|_{V_\odot}^2)^{1/2}\|q_\odot\|_{Q_\odot}.
\end{aligned}$$

Next, the inf-sup condition reads

$$\sup_{[u,u_\odot]\in[V,V_\odot]} \frac{\varepsilon\langle\overline{\mathcal{T}}_\Lambda u - u_\odot, q_\odot\rangle_\Lambda}{(\|u\|^2_{\sqrt{\kappa}H^1(\Omega)} + \varepsilon^2\|u_\odot\|^2_{\sqrt{\kappa_\odot}H^1(\Lambda)})^{1/2}} \geq \beta\|q_\odot\|_{Q_\odot}.$$

We first note that $(\alpha X)' = \alpha^{-1}X'$ holds for the dual of weighted Hilbert space. In turn, let R_1 be the Riesz map from $\kappa_\odot^{-1/2}H^{-1}$ to $\kappa_\odot^{1/2}H^1$ and $R_{1/2}$ be the Riesz map from $\kappa^{-1/2}H^{-1/2}$ to $\kappa^{1/2}H^{1/2}$. Then $\hat{u}_\odot = -\frac{1}{\varepsilon}R_1 q_\odot$ implies that $\varepsilon\|\hat{u}_\odot\|_{\sqrt{\kappa_\odot}H^1} = \|q_\odot\|_{\kappa_\odot^{-1/2}H^{-1}}$. From [29], we also see that there exists a harmonic extension operator H such that for $\hat{u} = H\hat{q}_\odot$, $\hat{q}_\odot \in H^{1/2}$. Letting $\hat{q}_\odot = R_{1/2}q_\odot$ we obtain $\hat{q}_\odot = \overline{\mathcal{T}}_\Lambda\hat{u}$ where $\|\hat{q}_\odot\|_{\sqrt{\kappa}H^{1/2}} = \|q_\odot\|_{\kappa^{-1/2}H^{-1/2}}$ and

$$\|\hat{u}\|_{H^1(\Omega)} \leq C_{IT}\varepsilon\|\hat{q}_\odot\|_{H^{\frac{1}{2}}(\Gamma_\varepsilon)} = \varepsilon C_{IT}\|q_\odot\|_{H^{-\frac{1}{2}}(\Gamma_\varepsilon)}.$$

We note that the latter inequality, generally named the *inverse trace inequality*, involves the generic constant C_{IT} that may possibly depend on the domain Γ_ε and precisely on its cross section quantified by the (small) parameter ε. Hence, we obtain

$$\sup_{[u,u_\odot]\in[V,V_\odot]} \frac{\varepsilon\langle\overline{\mathcal{T}}_\Lambda u - u_\odot, q_\odot\rangle_\Lambda}{(\|u\|^2_{\sqrt{\kappa}H^1(\Omega)} + \varepsilon^2\|u_\odot\|^2_{\sqrt{\kappa_\odot}H^1(\Lambda)})^{1/2}}$$

$$\geq \frac{\varepsilon\langle \overline{\mathcal{T}}_\Lambda \hat{u} - \hat{u}_\odot, q_\odot\rangle_\Lambda}{(\|\hat{u}\|^2_{\sqrt{\kappa}H^1(\Omega)} + \varepsilon^2\|\hat{u}_\odot\|^2_{\sqrt{\kappa_\odot}H^1(\Lambda)})^{1/2}}$$

$$\geq \frac{\varepsilon(\langle R_{1/2}q_\odot, q_\odot\rangle + \langle\frac{1}{\varepsilon}R_1 q_\odot, q_\odot\rangle)}{(C_{IT}\varepsilon^2\|q_\odot\|^2_{\kappa^{-1/2}H^{-1/2}(\Lambda)} + \|q_\odot\|^2_{\kappa_\odot^{-1/2}H^{-1}(\Lambda)})^{1/2}} \geq \beta\|q_\odot\|_{Q_\odot}.$$

□

It should be noted that a proper choice of weighted-intersected Sobolev spaces has led to the removal of any *explicit* dependence of the inf-sup constant β on κ, $\kappa_\odot$ and ε. On the other hand, caution is needed as, through the possible dependence of C_{IT} on ε, the radius may influence β, and in turn the robustness of the estimates in ε.

2.2 Numerical Evidence About Preconditioning the Mixed-Dimensional Problems

The exploitation of iterative solvers, such as Krylov methods, is crucial to obtain fast solution algorithms for the discretized problems, yet, their performance essentially depends on the existence of a (practical) preconditioner which improves the spectral properties of the linear systems [44]. In constructing preconditioners for the *3d-1d*-problems, a main objective is to establish a parameter robustness preconditioner, i.e. a preconditioner with performance that does not deteriorate with respect to discretization parameters (mesh and element sizes), variations in material parameters ($\kappa > 0$, $\kappa_\odot > 0$) as well as the geometrical parameter ε, crucial for *3d-1d* problems. As an example, in the context of *3d-1d* microcirculation model (5), the radius of the vessel appears explicitly and implicitly in the *3d-1d* problem formulation, as the residue of the link between the three-dimensional topology and the one-dimensional one. Then it must be taken into account.

Here we wish to construct robust preconditioners for the coupled $3d$-$1d$ problem (6) which induces an operator equation: Find $x = [u, u_\odot, p_\odot] \in X$ such that

$$\mathcal{A}x = f \text{ in } X', \quad \mathcal{A} := \begin{pmatrix} -\kappa\Delta & & \varepsilon\overline{\mathcal{T}}_\Lambda{}' \\ & -\kappa_\odot\varepsilon^2\Delta_\Lambda & -\varepsilon I \\ \varepsilon\overline{\mathcal{T}}_\Lambda & -\varepsilon I & 0 \end{pmatrix}, \tag{9}$$

where, for a one-dimensional manifold as Λ, the operator Δ_Λ is equivalent to the second derivative in the direction tangent to Λ, previously denoted as d_s^2.

To establish preconditioners, we follow the framework of operator preconditioning [33]. That is, having shown stability of (6) we define the preconditioner as a Riesz map $\mathcal{B} : X' \to X$ with respect to inner products/norms in which the problem was shown to be well-posed. In turn, the framework relates the conditioning of the (preconditioned) operator $\mathcal{B}\mathcal{A} : X \to X$ to the stability constants of the Brezzi theory. More precisely, independence of the constants with respect to a particular problem parameter translates to robustness of the preconditioner (in the said parameter variations). Note that in case of (6), the critical Brezzi constants are C_B and β related to the boundedness and inf-sup condition of the bilinear form b, cf. Theorem 1. We remark that to obtain a robust discrete preconditioner, stable discretization is required in addition to well-posedness of the continuous problem. In particular, the Brezzi constants of the discretized problem must be independent of the discretization parameter.

Applying operator preconditioning, the analysis of well-posedness in [29] and Theorem 1 yields two different preconditioners for (9). Here, a key difference is the construction of the multiplier space, as [29] considers $H^{-1/2}(\Lambda)$. However, this space does not lead to robust algorithms as we shall demonstrate next. In fact, intersection spaces of Theorem 1 will be needed to obtain robustness in the material parameters. At the same time, at the end of this work it will become clear that the radius ε, as a parameter, plays a special and critical role in the robustness of a preconditioner originating from Theorem 1.

***2d-1d* Preconditioning Example**

Let us illustrate the challenges of developing a robust preconditioner for mixed-dimensional equations using a $2d$ analogue of (9) where we let $\Lambda = \Lambda_\varepsilon$ be a closed curved contained in the interior of $\Omega \subset \mathbb{R}^2$, cf. Fig. 2. Here, the coupling between the problem unknowns defined on Ω and Λ shall be realized by the standard $2d$-$1d$ trace operator leading to the coupled problem: Find $y = [u, u_\odot, p_\odot] \in Y$ such that

$$\mathcal{A}y = f \text{ in } Y', \quad \mathcal{A} = \begin{pmatrix} -\Delta & & \mathcal{T}_\Lambda{}' \\ & -\kappa_\odot\Delta_\Lambda & -I \\ \mathcal{T}_\Lambda & -I & 0 \end{pmatrix}. \tag{10}$$

Note that (10) is structurally similar to the $3d$-$1d$ coupled problem (9). However, for simplicity, the model $2d$-$1d$ problem contains only a single parameter, $\kappa_\odot > 0$, whose value can be (arbitrarily) large or small. We also recall that Λ is closed, and thus, in order to have an invertible operator on the whole of $H^1(\Lambda)$, we set $-\Delta_\Lambda = -\Delta_\Lambda + I$ in this section. We finally remark that in (9) the Laplacian on Λ is invertible due to boundary conditions imposed on $\partial\Omega_\ominus$ outside the coupling surface.

Following [27] where (10) has been shown to be well-posed with $Y = H^1_0(\Omega) \times H^1(\Lambda) \times H^{-1/2}(\Lambda)$ we shall construct preconditioners as Riesz maps with respect to two different inner products on Y leading to operators

$$\mathcal{B}_0 = \begin{pmatrix} -\Delta & & \\ & -\kappa_\odot\Delta_\Lambda & \\ & & -\Delta_\Lambda^{-1/2} \end{pmatrix}^{-1}, \quad \mathcal{B}_1 = \begin{pmatrix} -\Delta & & \\ & -\kappa_\odot\Delta_\Lambda & \\ & & -\Delta_\Lambda^{-1/2} - \kappa_\odot^{-1}\Delta_\Lambda^{-1} \end{pmatrix}^{-1}. \tag{11}$$

Here $\mathcal{B}_0$ stems from the analysis in [27] and can be seen to be analogous to the Riesz map preconditioner established in [29] for the coupled $3d$-$1d$ problem (9). In particular, the multiplier block of the preconditioner is a Riesz map of $H^{-1/2}(\Lambda)$ and is independent of the parameter $\kappa_\odot$. On the other hand, with $\mathcal{B}_1$ the multiplier is sought in the intersection space $H^{-1/2}(\Lambda) \cap \kappa_\odot^{-1/2} H^{-1}(\Lambda)$, cf. Theorem 1.

We shall compare the two preconditioners in terms of their spectral condition numbers defined as the ratio between the largest and smallest in magnitude eigenvalues of the problem: Find $y \in Y$, $\lambda \in \mathbb{R}$

$$\mathcal{A}y = \lambda \mathcal{B}_i^{-1} y \quad \text{in } Y'. \tag{12}$$

To assess robustness of the preconditioners we consider the generalized eigenvalue problem (12) for different values of $10^{-8} \leq \kappa_\odot \leq 10^8$ and refinements of the domain $\Omega = (-1, 1)^2$ with $\Gamma_\varepsilon = \{x \in \Omega, |x| = 0.1\}$. We then discretize the problem by continuous linear Lagrange elements ($\mathbb{P}_1$) where the finite element mesh of Ω always conforms to Λ, cf. Fig. 2.

Before addressing the preconditioners, let us briefly comment on the approximation property of the chosen discretization. To this end we note that (10) is associated with a system

$$\begin{aligned} -\Delta u &= f && \text{in } \Omega, \\ -\kappa_\odot \Delta_\Lambda u_\odot - [\![\nabla u]\!] \cdot \nu &= f_\odot && \text{on } \Lambda, \\ u - u_\odot &= g && \text{on } \Lambda \end{aligned} \tag{13}$$

which is here supplied with homogeneous Dirichlet boundary conditions for u, $u_\odot$ on $\partial\Omega$ and $\partial\Lambda$ respectively. Here, $[\![\cdot]\!]$ is the jump operator on Λ, $[\![v]\!] = v^+ - v^-$, defined with respect to the normal vector on the curve which points from the positive to the negative side.

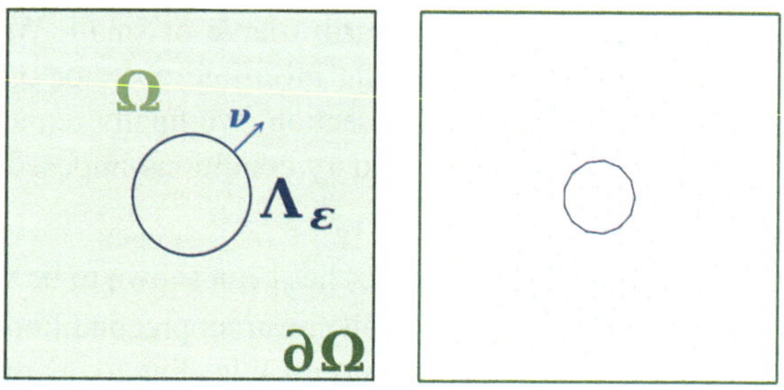

Fig. 2 (Left) Geometrical setting of the *2d-1d* problem (13). Domain Λ_ϵ plays similar role to the centerline in *3d-1d* system (5). (Right) In the experiments triangulation of Ω conforms to Λ

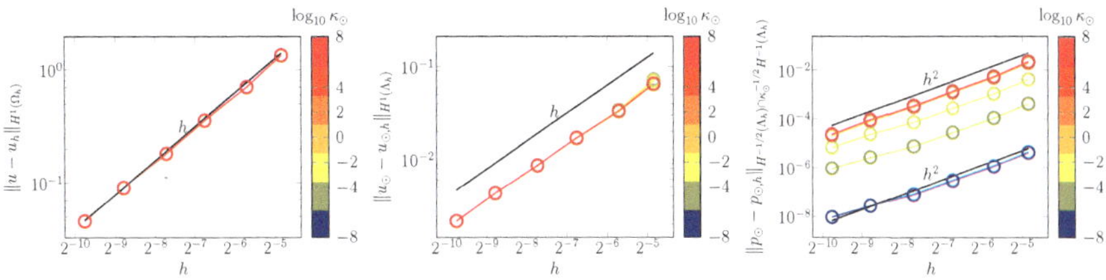

Fig. 3 Approximation errors of $\mathbb{P}_1$ discretization of (10) for different values of $\kappa_\odot$

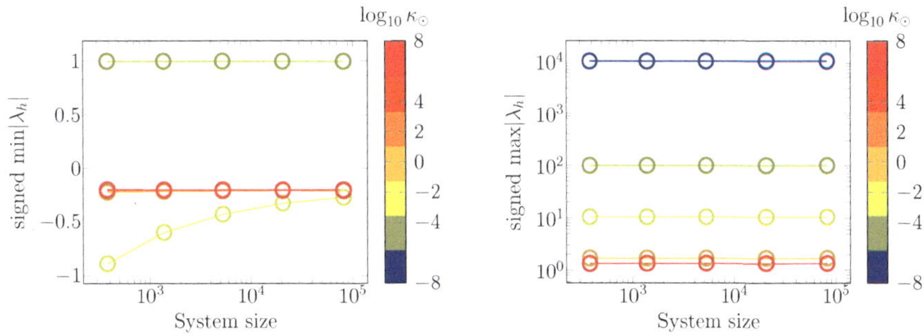

Fig. 4 Performance of preconditioner $\mathcal{B}_0$ from (11) for problem (10) with varying $\kappa_\odot$. Here $\Omega = (-1, 1)^2$, Γ_ε is a circle of radius $\varepsilon = 0.1$. All finite element spaces are discretized by $\mathbb{P}_1$ elements. The preconditioner does not produce $\kappa_\odot$-bounded condition numbers

Using (13) we measure the convergence of the discrete approximations u_h, $u_{\odot,h}$, $p_{\odot,h}$ in the norms induced by the (symmetric and positive definite) operator $\mathcal{B}_1^{-1}$. Here, the linear systems stemming from discretization[1] by $\mathbb{P}_1$ are solved with a preconditioned MinRes solver using the relative tolerance of 10^{-10}, see also Fig. 6.

The approximation errors obtained are plotted in Fig. 3, which, for all the values of $\kappa_\odot$, shows linear convergence in the respective H^1-norms for the error in u and $u_\odot$. Quadratic convergence can be seen for the multiplier in the norm of the intersection space $H^{-1/2}(\Lambda) \cap \kappa_\odot^{-1/2} H^{-1}(\Lambda)$. Without including the results, we remark that the intersection norm is essential for obtaining robust approximation. In particular, we observed that measuring convergence in only the $H^{-1/2}$-norm yields quadratic convergence for large values of $\kappa_\odot$ while for small values the rate drops to linear.

Having verified the discretization scheme (and our implementation), we return to preconditioning and stability of the eigenvalue problem (12). We summarize the results in Figs. 4 and 5 that show the extrema, that is, the minimum and maximum absolute values of the eigenvalues λ_h of the discretized problem (12). We note that

[1] In all the presented numerical examples FEniCS [32]-based module FEniCS$_{\text{ii}}$ [26] was used to discretize the coupled problems.

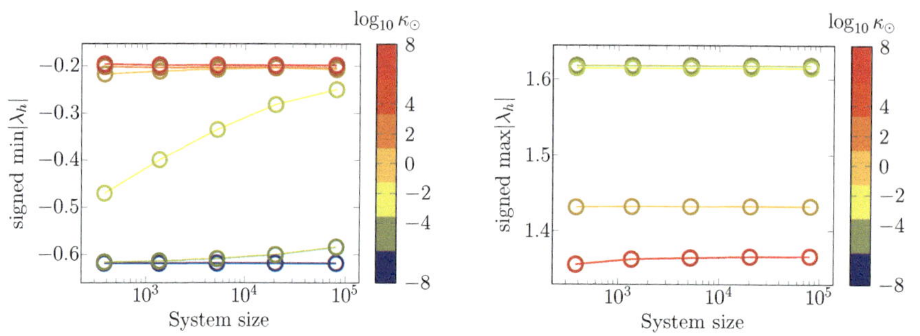

Fig. 5 Performance of preconditioner $\mathcal{B}_1$ from (11) for $2d$-$1d$ coupled problem (10) with varying $\kappa_\odot$. Setup as in Fig. 4. The preconditioner is robust in $\kappa_\odot$ and in mesh size

the bounds are plotted together with their sign which carries relevant information related e.g. to the discrete inf-sup or coercivity conditions.

Using $\mathcal{B}_0$, the extremal eigenvalues of (12) are displayed in Fig. 4. We observe that for each $\kappa_\odot$ the quantities are bounded in h verifying the stability of the discrete problem with $\mathbb{P}_1$ elements and the norm induced on Y by $\mathcal{B}_0^{-1}$, see [27]. However, there is an apparent growth of the largest eigenvalue in magnitude as $\kappa_\odot$ becomes small. As such, $\mathcal{B}_0$ does not produce a robust parameter solver. On the other hand, the preconditioner $\mathcal{B}_1$ yields spectral bounds that are stable in $\kappa_\odot$ and mesh refinement. We note that for small $\kappa_\odot$ the observed upper and lower bounds are close to their theoretical values [36, 43] of $(1+\sqrt{5})/2$ and $(1-\sqrt{5})/2$ respectively. This is due to the multiplier norm being then dominated by the matrix of the H^{-1}-inner product such that the discrete preconditioner (which computes $-\Delta_\Lambda^{-1}$ by LU decomposition) is close to being the exact Schur complement preconditioner.

To illustrate how conditioning translates into performance of iterative solvers, Fig. 6 reports the iteration counts of the MinRes solver using the two preconditioners (11). Here, we reuse the setup of the previous eigenvalue experiments while the right-hand side f in (10) is based on the manufactured solution setup described above. For each value of $\kappa_\odot$ and the mesh size, the solver starts from 0 initial guess and ends once the preconditioned residual norm is reduced by factor 10^{10}. Both preconditioners are computed exactly using LU for the two leading blocks, while the multiplier block, in particular the fractional term $-\Delta_\Lambda^{-1/2}$ is realized via spectral decomposition,[2] see [27] for a precise definition.

In Fig. 6 it can be seen that for large values of $\kappa_\odot$ both preconditioners yield iterations which are stable in mesh refinement. However, for small values, that is, when the term $\kappa_\odot^{-1}\Delta_\Lambda^{-1}$ becomes large, $\mathcal{B}_0$ shows a dependence on the parameter.

[2] Spectral decomposition is not suitable for practical applications because of its cubic scaling. However, Riesz maps at intersections of fractional-order Sobolev spaces can be approximated with optimal complexity by rational approximations [7].

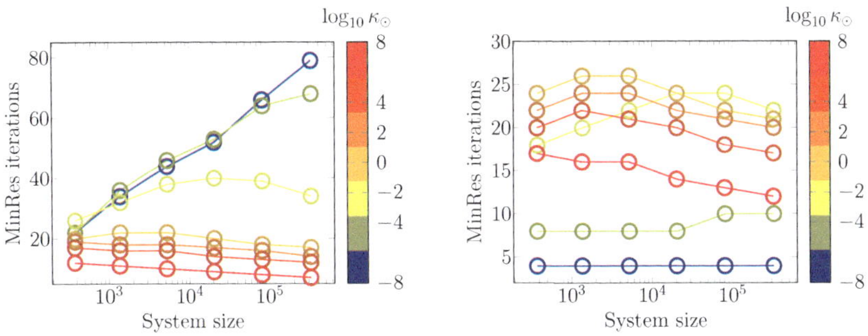

Fig. 6 Convergence of (11)-preconditioned MinRes solver for the problem (10) with varying $\kappa_{\odot}$. Only the preconditioner $\mathcal{B}_1$ (right) reflecting $p_{\odot} \in H^{-1/2} \cap \kappa_{\odot}^{-1/2} H^{-1}$ is parameter robust. Setup as in Fig. 4

We conclude that the intersection space exploited in the definition of $\mathcal{B}_1$ is crucial for parameter robustness.

3 Definition of a Preconditioner for the *3d-1d* Problem: Performances and Drawbacks

At this point let us return to the original coupled $3d$-$1d$ problem (9) that we shall now consider with a preconditioner

$$\mathcal{B} = \begin{pmatrix} -\kappa\Delta & & \\ & -\kappa_{\odot}\varepsilon^2\Delta_{\Lambda} & \\ & & -\frac{\varepsilon^2}{\kappa}\Delta_{\Lambda}^{-1/2} - \kappa_{\odot}^{-1}\Delta_{\Lambda}^{-1} \end{pmatrix}^{-1}, \quad (14)$$

i.e. the Riesz operator associated to the inner product of the space in which well-posedness of (9) was shown in Theorem 1. We note that for the approximate inversion of the diagonal blocks of (14), optimal algorithms are widely recognized in numerous instances, such as multigrid, domain decomposition, and Fast Fourier Transform (FFT) methods, particularly for standard Riesz maps in H^1, $H(\text{div})$, $H(\text{curl})$, etc. Given the widespread recognition of these methods in the field, providing references might be superfluous as these concepts are generally well established and well known. Moreover, recent developments have extended optimal algorithms to encompass fractional operators, broadening the applicability and efficiency of these numerical strategies [2, 5, 8, 16].

In order to test (14) we consider cylindrical domains $\Omega_{\oplus}$, $\Omega_{\ominus}$ each with height 1 and radii of 0.5 and ε respectively. Following [29] we discretize $\Omega = \Omega_{\oplus} \cup \Omega_{\ominus}$ so that the mesh conforms both to the interface Γ_{ε} and the centerline Λ, see Fig. 7. We remark that the conformity assumption was used in [29] to show stability of the

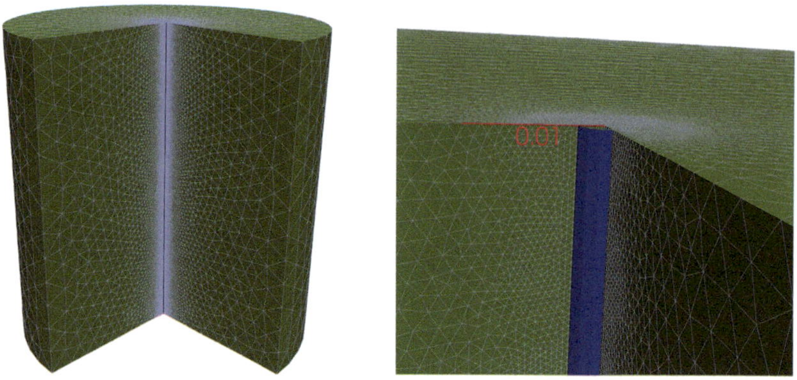

Fig. 7 Computational mesh considered in experiments for the coupled $3d$-$1d$ problem (14). Here $\varepsilon = 2 \cdot 10^{-3}$. The mesh of Ω conforms to the centerline Λ and to the coupling surface Γ_ε leading to high refinement near Λ (the edges of the mesh elements are colored light blue). (Right) The virtual surface used to calculate $\overline{\mathcal{T}}_\Lambda$ is rendered in dark blue color

discrete problem with $\mathbb{P}_1$ elements (used below). At the same time, the assumption leads to greatly refined meshes in the vicinity of Γ_ε. It also increases the cost of mesh generation and, due to the size of the resulting system, restricts[3] the type of experiments we can perform (on our serial computational setup). In the following we shall thus limit the computational study only to robustness of iterative methods. We note that stable discretization of (14) is also possible if the mesh of Ω is independent of Λ and Γ, cf. [29]. However, we argue that the conforming setup is simpler and more transparent.

In Fig. 8 we report on the convergence of the MinRes solver using the preconditioner (14) for the $2 \cdot 10^{-3} \leq \varepsilon \leq 10^{-1}$ and two-parameter regimes. Here, the action of the leading block $-\kappa\Delta^{-1}$ of (14) is approximated in terms of (i) a single V-cycle of algebraic multigrid (AMG) and (ii) 10 steps of the preconditioned conjugate gradient (PCG) method using AMG as preconditioner. The remaining two blocks of (14) are computed by LU factorization. We remark that choice (i) is more practical, while with (ii) the preconditioner is almost exact as the absolute residual norm after the 10 PCG steps is typically $< 10^{-15}$ in our case. With the case (ii) we thus aim to ensure that the effects of parameter variations on MinRes convergence are (mostly) due to the construction of the multiplier preconditioner. In both cases, the convergence criterion for the MinRes solver requires reducing the preconditioned residual norm by a factor 10^{10}. Finally, the coupling operator $\overline{\mathcal{T}}_\Lambda$ is approximated using a Legendre quadrature of degree 20.

[3] As an example, for radius $\varepsilon = 1 \cdot 10^{-2}$ the finest mesh considered contained roughly 11 million tetrahedra. Using $\mathbb{P}_1$ elements, the number of $3d$ unknowns is then ~2 million, while the multiplier space has ~2 thousand degrees of freedom.

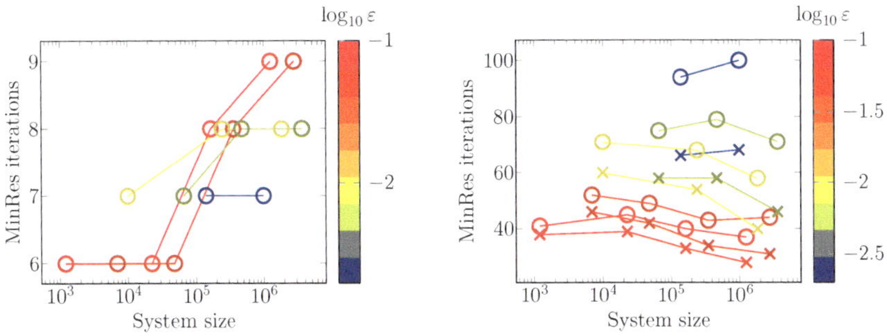

Fig. 8 Performance of preconditioner (14) for the coupled $3d$-$1d$ problem (9) and varying radius. (Left) We set $\kappa = 10^8$, $\kappa_\odot = 10^{-8}$. (Right) We set $\kappa = 10^{-8}$, $\kappa_\odot = 10^8$. The leading block of the preconditioner uses single AMG V-cycle (∘ markers) or 10 PCG iterations with AMG preconditioner (× markers). The remaining blocks are realized by LU

Taking into account the results in Fig. 8 we observe that the performance of the preconditioner differs dramatically between the two regimes. When $\kappa = 10^8$, $\kappa_\odot = 10^{-8}$, such that the H^{-1}-term can dominate the multiplier block in $\mathcal{B}$, the iteration counts are practically independent of ε. Note that here only the construction (i) for the leading block is considered, as it already yields low enough iterations. On the other hand, with $\kappa = 10^{-8}$, $\kappa_\odot = 10^8$ the solver performance deteriorates for small radii. This is true for the construction (i) which, for the different radii $\varepsilon \in \{1 \cdot 10^{-1}, 5 \cdot 10^{-2}, 1 \cdot 10^{-2}, 5 \cdot 10^{-3}, 2 \cdot 10^{-3}\}$ and the finest refinement levels, yields the iteration counts of 37, 44, 58, 71, 100 as well as for (ii) where convergence is reached respectively after 28, 31, 40, 46, 68 iterations.

The unbounded iterations, in particular with preconditioner (ii), bring into question the stability of the coupled $3d$-$1d$ problem (9) with preconditioner (14) and in particular the intersection space $Q_\odot$ for the multiplier. The lack of robustness is surprising, as it suggests that radius ε in (9) does not behave as a standard material parameter in the sense that the corresponding weighting in the (appropriate) intersection space does not yield robustness with respect to its variations. However, this result may be qualitatively justified on the basis of stability analysis, where we noticed the influence of the constant C_{IT} on the inf-sup condition. In the next section, we will show that C_{IT} may depend on ε, which may explain the findings in Fig. 8.

Varying Radius in *2d-1d* Preconditioning Example

In order to gain more insight into the observation that in the $3d$-$1d$ setting of (9) the multiplier preconditioner reflecting the intersection space $\kappa^{-1/2}\varepsilon H^{-1/2} \cap \kappa_\odot^{-1/2} H^{-1}$ did not lead to robustness, let us return to the $2d$-$1d$ operator (10). Next, we investigate the properties of the preconditioner $\mathcal{B}_1$ when ε varies. We recall that $\mathcal{B}_1$ was found to be robust with respect to the material parameters.

Using the previous experimental setup, Fig. 9 shows the performance of the preconditioner when $\kappa_\odot = 10^{10}$ and $\varepsilon \leq 10^{-1}$. Here, by choosing a large $\kappa_\odot$

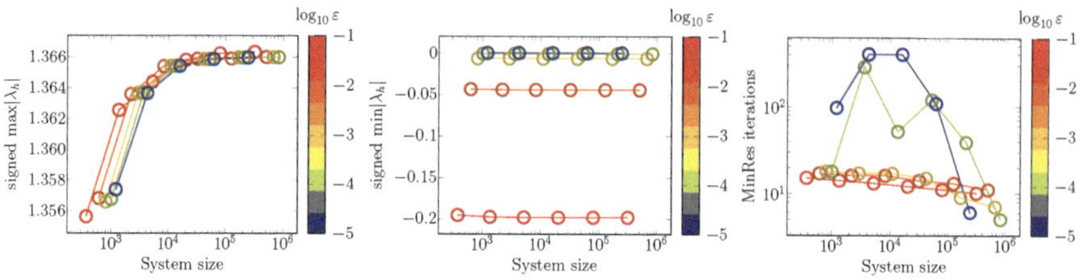

Fig. 9 Performance of preconditioner $\mathcal{B}_1$ in (11) for problem (10) with $\kappa_\odot = 10^{10}$, $\Omega = (-1, 1)^2$, Γ_ε is a circle of radius $10^{-5} \le \varepsilon \le 10^{-1}$. All finite element spaces are discretized by $\mathbb{P}_1$ elements. (Left, center) The conditioning deteriorates with ε as the largest in magnitude eigenvalue (12) is bounded in the parameter while the smallest in magnitude eigenvalue decreases with ε. (Right) For small radii the iterations are not stable (the maximum number of allowed iterations is set to 400)

we wish to put emphasis on the fractional part of the multiplier preconditioner, which, analogously to Theorem 1, brings the inverse trace constant into the Brezzi estimates. In Fig. 9 we observe that while the largest eigenvalues are bounded in ε, the smallest in magnitude eigenvalue approaches 0 as Λ_ε shrinks. We note that for all ε the two eigenvalue bounds are stable in mesh size h. In the figure, we further report convergence history of the MinRes solver (run with the same settings as in the previous experiments). We observe that for small radii the iteration counts grow rapidly with initial mesh refinement before decreasing back on finest meshes, cf. bounded iterations with mesh refinement in Fig. 6. However, the limit $\varepsilon \to 0$ does not seem to affect the iterative solver as clearly as blow-up of the condition number, cf. min$|\lambda_h|$ in Fig. 9. We attribute this behavior to the specific choice of the right-hand side and 0 initial guess in our numerical setup. However, the sensitivity to the radius furthers our claim of the special role of ε in understanding the robustness of the preconditioner.

4 The Role of the Inner Radius on Mixed-Dimensional Problems

As discussed in the previous sections, it has been observed that the robustness of the preconditioner (14) decreases as the inner radius of the inclusion approaches zero. This behavior is not limited to the $3d$-$1d$ preconditioner formulation but is also evident in the $2d$-$1d$ examples. Consequently, it suggests that the detrimental effect of ε is not the result of the dimensional reduction technique itself, but rather inherent in the mathematical structure of the problem.

Specifically, the dependence on ε is manifested in the constant of the inf-sup stability property. The analysis in Theorem 1 reveals that the boundedness constant C_B of the bilinear form b remains independent of the inner diameter. However, the inf-sup constant β is not guaranteed to be independent of ε as the presence of the inverse trace constant C_{IT} introduces questions regarding its independence from the inner radius. The inverse trace theorem, a well-established result in functional analysis, provides insight into this matter (see, for example, [31, 37, 46]). However, deriving an accurate bound for C_{IT} is a challenging task, particularly when considering a parameterized domain.

In this section, our objective is to investigate the relationship between the inf-sup constant β and the parameter ε, which is closely associated with the inverse trace constant C_{IT}. To accomplish this, we analyze a simplified yet significant scenario that allows us to elucidate the connection between the inf-sup constant and the inner radius.

4.1 The 2d-1d Formulation for the Perforated Domain Problem

Consider a generalized cylindrical vessel immersed in a three-dimensional domain Ω. The surface of the vessel is indicated by Γ_ε, where ε indicates the radius of the cylinder. We are interested in studying how the mathematical structure of the problem behaves when $\varepsilon \to 0$. A straightforward approach can be to select a slice of Ω, here denoted by the superscript "$*$", such that we obtain a domain Ω^*, where the curve Γ_ε^* is the restriction of Γ_ε on the slice, see Fig. 10. Clearly, $\text{diam}\Gamma_\varepsilon^*$ depends on ε. In turn, studying the effects of shrinking $\text{diam}\Gamma_\varepsilon^*$ in Ω^* (a $2d$-$1d$ system) is representative of the effects of the decreasing vessel / inner radius.

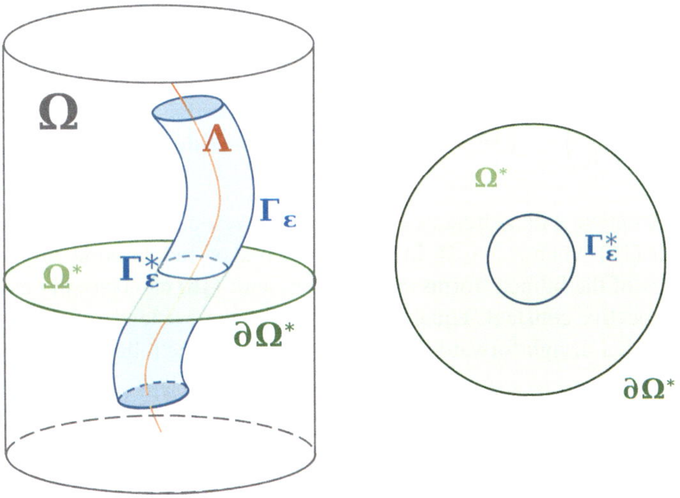

Fig. 10 (Left) Pictorial representation of a vessel with radius ε, centerline Λ and boundary Γ_ε immersed in a three-dimensional domain Ω. By cutting a slice of Ω, we obtain the two-dimensional domain Ω^* with boundary $\partial\Omega^*$ and circular inclusion in one-dimensional Γ_ε^*. (Right) Pictorial representation of the slice domain Ω^*

The underlying assumption is the complete decoupling of the radius influence from the axial direction, which seems reasonable when considering a cylindrical setting. Indeed, inspecting the expression of the Laplacian in cylindrical coordinates

$$\Delta = \frac{1}{\rho}\frac{\partial}{\partial\rho}\left(\rho\frac{\partial}{\partial\rho}\right) + \frac{1}{\rho^2}\frac{\partial^2}{\partial\phi^2} + \frac{\partial^2}{\partial z^2}$$

corroborates the assumption: ρ (and in turn ε) does not appear in the axial derivative part. From now until the end of this section, the superscript "$*$" will be omitted, so that Ω and $\partial\Omega$ will refer respectively to the $2d$ sliced domain and its outer boundary. Moreover, Γ_ε will denote the closed one-dimensional curve embedded in Ω^*.

Let us consider the following ε-dependent Poisson problem defined in Ω:

$$\begin{aligned} -\Delta u &= 0 \quad &&\text{on } \Omega,\\ u|_{\Gamma_\varepsilon} &= g \quad &&\text{on } \Gamma_\varepsilon,\\ u &= 0 \quad &&\text{on } \partial\Omega, \end{aligned} \tag{15}$$

where $\Gamma_\varepsilon \cap \partial\Omega = \emptyset$ and u is a two-dimensional scalar field defined on Ω. Now consider a mixed weak formulation of (15), where the boundary condition on Γ_ε is weakly enforced by a Lagrange multiplier to allow the mathematical structure of (15) to mirror the $3d$-$1d$ problem. The variational problem reads:

$$\begin{aligned} (\nabla u, \nabla v)_\Omega + (p_\odot, \mathcal{T}v)_{\Gamma_\varepsilon} &= 0 \qquad && \forall u, v \in H_0^1(\Omega)\\ (\mathcal{T}u, q_\odot)_{\Gamma_\varepsilon} &= (g, q_\odot)_{\Gamma_\varepsilon} \qquad && \forall p_\odot, q_\odot \in H^{-1/2}(\Gamma_\varepsilon) \end{aligned}. \tag{16}$$

Note that the solution operator $\mathcal{A}$ of (16) has the following block structure

$$\mathcal{A} = \begin{pmatrix} A & B' \\ B & 0 \end{pmatrix}, \quad \langle Au, v\rangle_\Omega = \int_\Omega \nabla u \cdot \nabla v, \quad \langle Bu, q_\odot\rangle_{\Gamma_\varepsilon} = \int_{\Gamma_\varepsilon} p_\odot v.$$

The question we will address next is that of the well-posedness of the variational formulation (16) when $\varepsilon \to 0$. In the framework of saddle-point problems, the boundedness of the bilinear forms a, b in the case of (16) can be easily established with the respective constants equal to 1. Regarding the existence of the Brezzi inf-sup constant β, a straightforward proof will make use of the following theorem [46]:

Theorem 2 (Inverse Trace Theorem) *Given the necessary smoothness assumption for Ω, the trace operator $\mathcal{T} : H^1(\Omega) \to H^{1/2}(\Gamma_\varepsilon)$ has a continuous right inverse operator*

$$\mathcal{E} : H^{1/2}(\Gamma_\varepsilon) \to H^1(\Omega)$$

satisfying $\mathcal{T}\mathcal{E}w_{\odot} = w_{\odot}$ *for all* $w_{\odot} \in H^{1/2}(\Gamma_\varepsilon)$ *as well as*

$$\|\mathcal{E}w_{\odot}\|_{H^1(\Omega)} \leq C_{IT}\|w_{\odot}\|_{H^{1/2}(\Gamma_\varepsilon)} \quad \forall w_{\odot} \in H^{1/2}(\Gamma_\varepsilon).$$

Then, by taking $v = \mathcal{E}v_{q_{\odot}}$, where, by $v_{q_{\odot}}$, we intend an element of $H^{1/2}(\Gamma_\varepsilon)$ such that the following Riesz mapping properties hold

$$\langle v_{q_{\odot}}, w_{\odot}\rangle_{H^{1/2}(\Gamma_\varepsilon)} = (q_{\odot}, w_{\odot})_{\Gamma_\varepsilon} \text{ and } \|v_{q_{\odot}}\|_{H^{1/2}(\Gamma_\varepsilon)} = \|q_{\odot}\|_{H^{-1/2}(\Gamma_\varepsilon)},$$

we obtain that

$$\begin{aligned}\sup_{v\in H_0^1(\Omega), v\neq 0} \frac{(\mathcal{T}v, q_{\odot})_{\Gamma_\varepsilon}}{\|v\|_{H_0^1(\Omega)}} &\geq \frac{(v_{q_{\odot}}, q_{\odot})_{\Gamma_\varepsilon}}{\|\mathcal{E}v_{q_{\odot}}\|_{H_0^1(\Omega)}} \\ &= \frac{\langle v_{q_{\odot}}, v_{q_{\odot}}\rangle_{H^{1/2}(\Gamma_\varepsilon)}}{\|\mathcal{E}v_{q_{\odot}}\|_{H_0^1(\Omega)}} \geq \frac{1}{C_{IT}}\|v_{q_{\odot}}\|_{H^{1/2}(\Gamma_\varepsilon)}.\end{aligned}$$

What is apparent from the above calculation is, again, the structural relationship between the existence of the inf-sup constant β, and the inverse trace inequality constant C_{IT} i.e $\beta = 1/C_{IT}$. This suggests (and is what we want to uncover) a tight relationship between β and the inner radius using the *trace inequality constant* C_T. Indeed, from Lemma 2.2 in [25], the influence of ε is made clear:

Lemma 1 (Lemma 2.2 in [25]) *Let* $B_\varepsilon \subset \mathbb{R}^2$ *be a circle with a sufficiently small radius* ε *and* $v \in H_0^1(\Omega)$. *Then we have:*

$$\|\boldsymbol{T}v\|_{L^2(\partial B_\varepsilon)} \leq C_T\,\|v\|_{H_0^1(\Omega)} = C\sqrt{\varepsilon|\log\varepsilon|}\,\|v\|_{H_0^1(\Omega)} \tag{17}$$

with C positive and independent of ε.

We remark that similar result to Lemma 1 holds also in the 3d-1d setting of problem (9), see Lemma 6.4 in [34].

Considering that the extension operator, $\mathcal{E}$, is the right inverse of the trace, $\boldsymbol{T}$, the following claim is proposed: *the radius of the vessel* ε *directly affects the inf-sup constant* β *through the constant of the inequality of the trace* C_T. This claim sounds reasonable if one considers that β directly characterizes the lower bound of the bilinear form b, which strongly depends on the trace operator; moreover, it is corroborated by the fact that $\beta = 1/C_{IT}$ so that, assuming a relation between the trace inequality and the inverse trace one, we find that ε directly affects β. However, to prove the claim, it would be required to track in detail the exact expression of C_{IT}, thus obtaining a direct link between β and ε. Establishing such an analytical expression for C_{IT} can be a very intricate task, especially when the domain size needs to be considered as a parameter. Nevertheless, we can rely on numerical analysis and employ the computability of the constants in a discretized setting, as follows.

Brezzi Inf-Sup Constant Evaluation Let $V = H_0^1(\Omega)$, $Q = H^{-1/2}(\Gamma_\varepsilon)$ and $\|\cdot\|_V$ be the H_0^1-norm (induced by the operator A) while $\|\cdot\|_Q$ shall be the $H_\Gamma^{-1/2}$-norm induced by the fractional operator $N := (-\Delta + I)^{-1/2}$ (where both terms are understood to be defined in Γ_ε). Consider now a family (V_h, Q_h) of discretization of (V, Q) characterized by the following discrete Brezzi inf-sup constant.

$$\sup_{v \in V_h, v \neq 0} \frac{b(v, q_\odot)}{\|v\|_{V_h}} \geq \beta_h \|q_\odot\|_{Q_h} \quad \forall q_\odot \in Q_h. \tag{18}$$

If (V_h, Q_h) is a *stable* discretization i.e. both the discrete Brezzi inf-sup and coercivity constants are bounded from below by a constant which is independent from h [1], [42]:

$$\{\beta_h\}_{h \to 0} \geq \beta > 0, \tag{19}$$

then an approximation of β can be obtained by considering the truncated limit $\{\beta_h\}_{h \to \delta}$, for δ sufficiently small. For the purpose of computing β_h, we recall the following lemma of Qin [41]:

Lemma 2 (Qin [41]) *Given a stable discretization* (V_h, Q_h) *for the saddle point problem* (16), *consider the following generalized eigenvalue problem: Find* $\lambda \in \mathbb{R}$, $0 \neq (u, p_\odot) \in V_h \times Q_h$ *such that*

$$\langle u_h, v \rangle_V + b(v, p_\odot) + b(u_h, q_\odot) = -\lambda \langle p_\odot, q_\odot \rangle_Q \quad \forall (v, q_\odot) \in V_h \times Q_h. \tag{20}$$

Then, $\lambda \geq 0$ *and* $\beta \approx \beta_h = \sqrt{\lambda^{\min}}$.

In the framework of the problem (16), the eigenvalue problem can be resolved in the following form involving the Schur complement of $\mathcal{A}$: Find $p_\odot \in Q_h$, $\lambda_B > 0$ such that[4]

$$BA^{-1}B' p_\odot = \lambda_B^2 N p_\odot \quad \text{in } Q_h'. \tag{21}$$

Here, the subscript B, which stands for "Brezzi", has been introduced for clarity of notation, as will become clear soon. In particular, for stability of the discrete problem (16), $\lambda_B^{\min} = \min \lambda_B$, $\lambda_B^{\max} = \max \lambda_B$ shall be independent of the mesh size h. We remark that in this framework $\lambda_B^{\min} \approx \beta$.

[4] Though it is λ_B^2 that is the eigenvalue of (21) we shall in the following, with a slight abuse of notation, refer to $\max \lambda_B = \lambda_B^{\max}$, $\min \lambda_B = \lambda_B^{\min}$ as the eigenvalue bounds, extremal eigenvalues or simply eigenvalues. This choice is intended to simplify the notation and avoid proliferation of $\sqrt{}$ in the text. A similar convention will be applied also to the Steklov eigenvalue problems (22) and (26).

Trace Constant Evaluation Let us consider the eigenvalue problem defined on the domain in Fig. 10, which reads as follows:

$$\begin{aligned} -\Delta u &= 0 && \text{in } \Omega \subset \mathbb{R}^2, \\ \nabla u \cdot \nu &= \lambda_S^{-2} u && \text{on } \Gamma_\varepsilon, \\ u &= 0 && \text{on } \partial\Omega, \end{aligned} \tag{22}$$

where ν is the unit normal to Γ_ε. We remark that (22) is a variant of the Steklov eigenvalue problem. For rigorous mathematical treatment of (22) we refer to e.g. [20] and references therein.

The weak formulation of (22), leads to the generalized eigenvalue problem: Find $u \in H_0^1(\Omega)$, $\lambda_S > 0$ satisfying

$$\int_{\Gamma_\varepsilon} uv = \lambda_S^2 \int_\Omega \nabla u \cdot \nabla v \quad \forall v \in H_0^1 \tag{23}$$

so that, $\lambda_S^{\max}$ allows for estimates of the $L^2(\Gamma_\varepsilon)$-norm of u in terms of the $H_0^1(\Omega)$-norm. Indeed, we have $\|u\|_{L^2(\Gamma_\varepsilon)} \le \lambda_S^{\max} \|u\|_{H_0^1}(\Omega)$ for all $u \in H_0^1(\Omega)$.

Once the eigenvalue problems (20) and (22) are properly discretized (from now on, the subscript h will denote the discretized analogues of a given quantity; subscript ε_i will denote a quantity evaluated on a domain with inner/inclusion radius equal to ε_i), the numerical tools for the evaluation of the inf-sup constant β and the trace inequality constant C_T are readily established. Therefore, we can proceed with the investigation of the influence of ε in the mathematical framework considered.

The research path is summarized in Fig. 11. Due to the analytical difficulty in the evaluation of the role of the inner radius in the inverse trace constant C_{IT}, and its relationship with the trace constant, we shift our inquiry to a discretized setting and the discrete eigenvalue problems. What will be done is a concomitant evaluation of $\{\beta_{h,\varepsilon_i}\}_{\varepsilon_i \in I}$ and $\{C_{T_{h,\varepsilon_i}}\}_{\varepsilon_i \in I}$, where $I = \{\varepsilon_0, \varepsilon_1, \ldots, \varepsilon_n\}$ and $\varepsilon_0 > \varepsilon_1 > \ldots > \varepsilon_n$ with $\varepsilon_n \ll 1$. Then, if there is some noticeable correlation between $\lambda_B^{\min}$ and $\lambda_S^{\max}$ exists, then it also holds for $\{\beta_{h,\varepsilon_i}\}$ and $\{C_{T_{h,\varepsilon_i}}\}$ (by means of (20) and (21)) so that we can find numerical evidence to support our claim.

4.2 Numerical Results About the 2*d*-1*d* Formulation

Here we summarize the numerical experiments concerning the evaluation of $\{\beta_{h,\varepsilon_i}\}$ and $\{C_{T_{h,\varepsilon_i}}\}$ with varying $\varepsilon_i \in I = \{10^{-1}, \ldots, 10^{-5}\}$. For simplicity we shall first limit the investigations to $\Omega = \{x \in \mathbb{R}^2, |x| < 1\}$, a circular inclusion $\Gamma_\varepsilon = \{x \in \Omega, |x| = \varepsilon\}$ and $\mathbb{P}_1$ discretization.

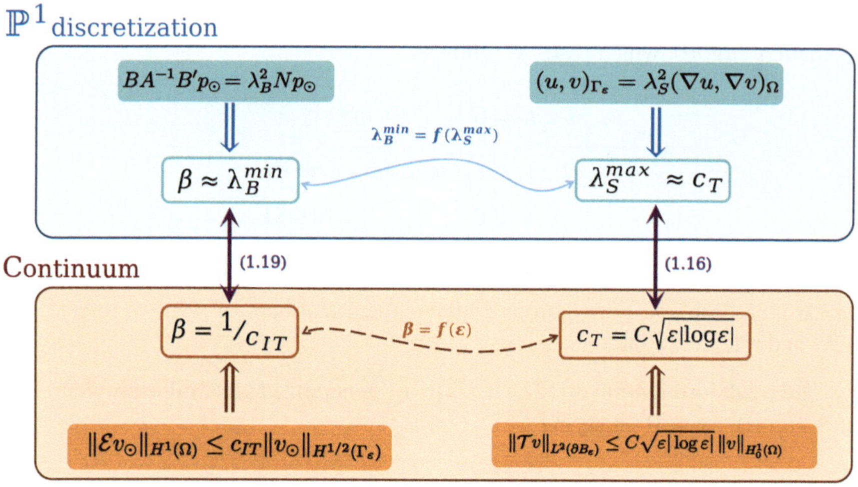

Fig. 11 Pictorial representation of the path followed in the analysis of the role of ε on β. Thanks to the evaluation of the eigenvalues of problems (21) and (22), a relation can be established between $\lambda_B^{\min}$ and $\lambda_S^{\max}$. Then, by virtue of Eqs. (17) and (23) (represented by the purple arrows) bridging the discrete setting to the continuous one, we can close the loop and determine a relation between the Brezzi inf-sup constant β and the inner radius ε

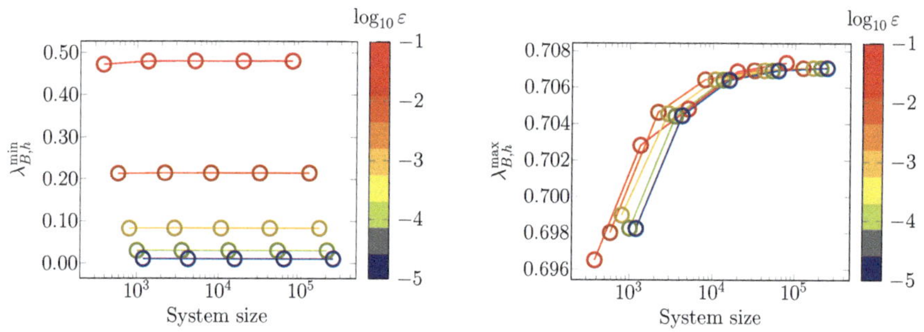

Fig. 12 (Left) Mesh convergence of the extremal eigenvalues of Schur complement (21) for $\Omega = \{x \in \mathbb{R}^2,\ |x| < 1\}$, $\Gamma_\varepsilon = \{x \in \Omega,\ |x| = \varepsilon\}$. Both spaces V and Q are discretized by $\mathbb{P}_1$ elements. For each ε a sequence of problems is considered on uniformly refined meshes starting from size $h_0 \geq h_l \geq h_{\min}$ leading to eigenvalues λ_{B,h_l}

In Fig. 12 we demonstrate that our choice of norms and finite element spaces yields a stable problem. More precisely, we observe that both the upper and lower bounds on the Schur complement spectra are stable in mesh refinement for each fixed ε. Moreover, $\lambda_B^{\max}$ appears to be bounded in the radius of Γ_ε, cf. Theorem 1. On the other hand, $\lambda_B^{\min}$ seems to decrease together with ε.

We now propose that $\lambda_B^{\min}$ is closely related to the Steklov eigenvalue problem (23). To investigate the relation between $\lambda_S^{\max}$ and $\lambda_B^{\min}$, Fig. 13 plots the relative

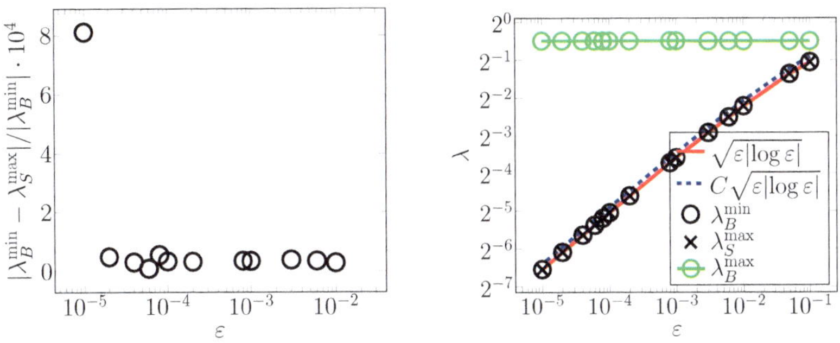

Fig. 13 (Left) Error between the smallest eigenvalue $\lambda_B^{\min}$ of the Schur complement problem (21) and the largest eigenvalue $\lambda_S^{\max}$ of the Steklov problem (23). In both cases, values from the finest level of refinement are considered, i.e. $\lambda_X := \lambda_{X,h_{\min}}$. (Right) Dependence of the Schur complement eigenvalues on the radius of coupling curve $\Gamma_\varepsilon = \{x \in \Omega, |x| = \varepsilon\}$. Value $C \approx 0.999$ is obtained by fitting values $\lambda_B^{\min}$ for $\varepsilon < 10^{-1}$

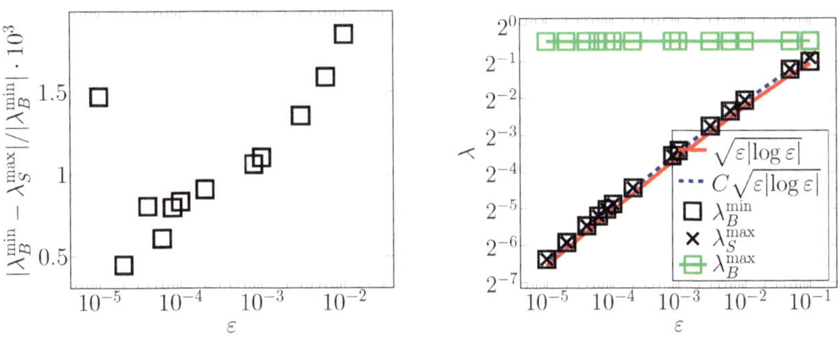

Fig. 14 (Left) Error between the smallest eigenvalue $\lambda_B^{\min}$ of the Schur complement problem (21) and the largest eigenvalue $\lambda_S^{\max}$ of the Steklov problem (23) on squared domain $\Omega = (-1, 1)^2$. In both cases, values from the finest level of refinement are considered, i.e. $\lambda_X := \lambda_{X,h_{\min}}$. (Right) Dependence of the Schur complement eigenvalues on the radius of coupling curve $\Gamma_\varepsilon = \partial(-\varepsilon, \varepsilon)^2$. Value $C \approx 1.116$ is obtained by fitting values $\lambda_B^{\min}$ for $\varepsilon < 10^{-1}$

error between the two quantities[5] for varying values of ε. With the relative error $\sim 10^{-4}$ it appears that $\lambda_S^{\max} = \lambda_B^{\min}$. In Fig. 13 we finally plot the dependence of $\lambda_S^{\max}$, $\lambda_B^{\min}$ on the radius of Γ_ε. The observed dependence agrees well with the theoretical bound [25].

To corroborate the independence of the established relation between $\lambda_B^{\min}$ and $\lambda_S^{\max}$ from geometrical factors, we carry out the above analysis for a square-shaped inclusion, i.e. $\Gamma_\varepsilon = \partial(-\varepsilon, \varepsilon)^2$ in a squared domain $\Omega = (-1, 1)^2$. A comparison between the obtained values of $\lambda_S^{\max}(\varepsilon)$ and $\lambda_B^{\min}(\varepsilon)$, is depicted in Fig. 14 (see also

[5] For each ε we consider a sequence of eigenvalues λ_{h_l} computed on meshes with sizes h_l. We terminate the sequence once the relative error between subsequent eigenvalues, $|\lambda_{h_l} - \lambda_{h_{l+1}}|/\lambda_{h_l}$, is less than 10^{-4}.

the Appendix for additional results). No significant differences are reported. The relative error between the eigenvalues remains well below the percentage point, and the expression $C\sqrt{\varepsilon|\log\varepsilon|}$ retraces $\lambda_B^{\min}(\varepsilon)$ with a constant $C \approx 1.116$.

Having shown independence of our observations from the shape of the inclusion/-coupling surface, the effect of the meshing strategy and the discretization of finite elements will be investigated using a circular embedded domain Γ_ε. To exclude any relevant influence of the mesh on the obtained results, the same analysis has been repeated on a specific type of mesh, which we shall refer to as *layered*, and which has the peculiarity of being Γ_{ε_i}-conformal (Ω vertices and edges are consistent with Γ_{ε_i} ones) for every value of $\varepsilon_i \in I$ simultaneously; cf. Fig. 15. In such a way, the mesh configuration, which is the same for every value of the inner radius, cannot be held responsible for any effects on λ-ε relations. The results are plotted in Fig. 16 (left). The relative difference between the values of $\lambda_B^{\max}(\varepsilon)$, $\lambda_B^{\min}(\varepsilon)$ and $\lambda_S^{\max}(\varepsilon)$ obtained

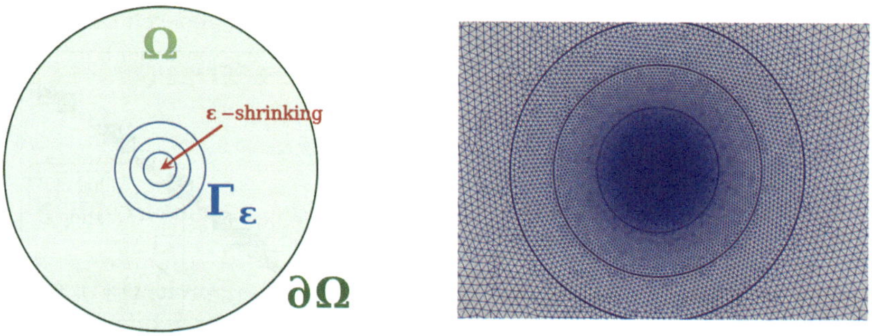

Fig. 15 (Left) Pictorial representation of the inner-radius shrinking process. (Right) Partial representation (zoomed toward the Ω center) of the *layered* mesh. The mesh has the property of being simultaneously conforming to every Γ_{ε_i}, $\varepsilon_i \in I$ (purple circles)

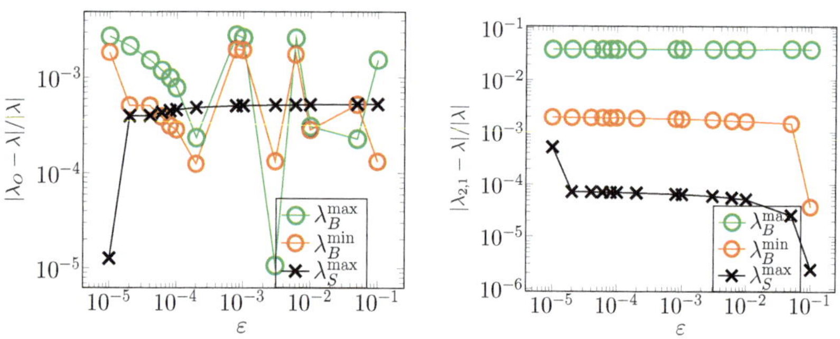

Fig. 16 Relative error $|\lambda_* - \lambda|/|\lambda|$, where λ spans $\{\lambda_B^{\max}, \lambda_B^{\min}, \lambda_S^{\max}\}$ eigenvalues obtained with standard meshing procedure (each grid conforming to a specific Γ_{ε_i}; see Fig. 13). (Left) $\lambda_* \equiv \lambda_O$ spans $\{\lambda_{B,O}^{\max}, \lambda_{B,O}^{\min}, \lambda_{S,O}^{\max}\}$ obtained on the *layered* mesh in Fig. 15 and denoted by the subscript "O". (right) $\lambda_* \equiv \lambda_{2,1}$ spans $\{\lambda_{B,2,1}^{\max}, \lambda_{B,2,1}^{\min}, \lambda_{S,2,1}^{\max}\}$ obtained with a $\mathbb{P}_2 - \mathbb{P}_1$ discretization and denoted by subscript "2, 1"

on different meshes (each conformal to a specific Γ_{ε_i}) and those obtained on the layered mesh differs by less than one percentage point. Consequently, the same relation as plotted in Fig. 13 between $\lambda_S^{\max}(\varepsilon)$ and $\lambda_B^{\min}(\varepsilon)$ also holds for the values obtained on the layered mesh. In conclusion, the results do not undergo significant grid influence. Additional numerical results in support of this conclusion can be found in Appendix.

In addition to the geometrical factors (shape, mesh), our aim is to exclude possible effects of different polynomial order between the spaces $2d$ and $1d$. The results of the comparison between the discretization $\mathbb{P}_1-\mathbb{P}_1$ and $\mathbb{P}_2-\mathbb{P}_1$ (continuous elements $\mathbb{P}_2$ for Ω and $\mathbb{P}_1$ for Γ_{ε_i}) are plotted in Fig. 16. No noticeable difference can be attributed to the change of polynomial order, except for the effects on $\lambda_B^{\max}(\varepsilon)$, which differ more than 1%, but still remain independent of ε_i. From this, we can deduce that also for the $\mathbb{P}_2 - \mathbb{P}_1$ discretization, $\lambda_B^{\min} \approx \lambda_S^{\max} \sim \sqrt{\varepsilon|\log\varepsilon|}$. Thus, no relevant role of the polynomial degree on the results can be experienced (for detailed results see the Appendix).

4.3 The 2d-0d Formulation for the Perforated Domain Problem

The link between stability of (15) and inner radius ε via the constant of the inverse trace inequality, which we demonstrated above for $2d$-$1d$ trace problem, naturally extends to $3d$-$2d$ coupling as well. In connection with the $3d$-$1d$ problem (6) we may ask if the dimensional reduction removes the observed effect of ε. To address this question we continue our investigations by applying model reduction to (15) leading (formally) to the system

$$\begin{aligned} -\Delta u &= 0 && \text{in } \Omega \subset \mathbb{R}^2, \\ \bar{u} &= g && \text{on } \Gamma, \\ u &= 0 && \text{on } \partial\Omega. \end{aligned} \tag{24}$$

We recall that $\bar{u}$ computes a mean of function $u : \Omega \to \mathbb{R}$ over the curve $\Gamma = \Gamma_\varepsilon$, i.e. $\bar{u} = |\Gamma|^{-1}\int_\Gamma u$. Note that through u and $\bar{u}$ the coupled problem (24) includes a dimensional gap of 2 (as in the case of the $3d$-$1d$ problem (6)).

Letting $V = H_0^1(\Omega)$ and $Q = \mathbb{R}$ the weak form of (24) read: Find $u \in V$, $p_\odot \in Q$ such that

$$\begin{aligned} \int_\Omega \nabla u \cdot \nabla v - \int_\Gamma p_\odot \bar{v} &= 0 \quad \forall v \in V, \\ -\int_\Gamma q_\odot \bar{u} &= 0 \quad \forall q_\odot \in Q. \end{aligned} \tag{25}$$

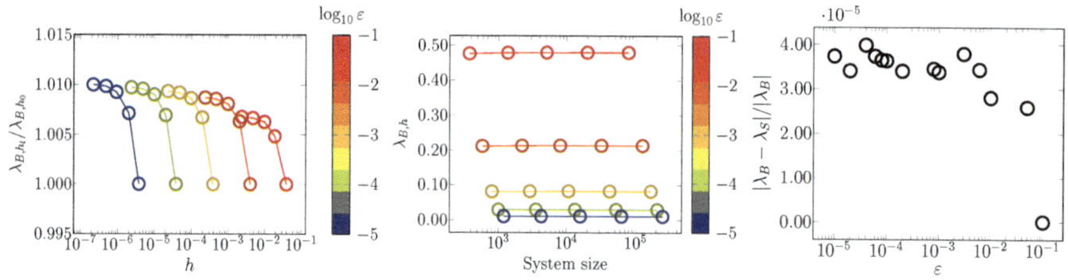

Fig. 17 (Left, center) Mesh convergence of the Schur complement eigenvalue of (25) for $\Omega = \{x \in \mathbb{R}^2, |x| < 1\}$, $\Gamma_\varepsilon = \{x \in \Omega, |x| = \epsilon\}$ and $\mathbb{P}_1$. For each ε a sequence of problems is considered in uniformly refined meshes starting from size $h_0 \geq h_l \geq h_{\min}$ leading to eigenvalues λ_{B,h_l}. (Right) Error between the Schur complement eigenvalues and the eigenvalues of the Steklov problem (26). In both cases, results for $h_l = h_{\min}$ are shown

For well-posedness of (25) when $\Gamma \subset \partial\Omega$ we refer to [15]. Here we shall consider Q with an inner product $(p_\odot, q_\odot)_Q = \int_\Gamma p_\odot q_\odot$ inducing the norm $\|p_\odot\|_Q = (p_\odot, p_\odot)_Q^{1/2}$. On V, the norm is given by the H^1-seminorm. From the point of Brezzi theory, a convenient property of the reduced problem is the fact that Q is one-dimensional. Thus, the Schur complement spectrum (21) contains only a single eigenvalue λ_B. Then, Fig. 17 shows computational evidence for the stability of (25) in $V \times Q$ with the chosen norms. Specifically, taking $\Omega = \{x \in \mathbb{R}^2 : |x| < 1\}$ such that the triangulation of the domain always conforms to Γ_ε, and using $\mathbb{P}_1$ elements, it can be seen that the Brezzi constant(s) related to the coupling operator[6] are bounded in the mesh size. However, similar to the full problem (15) we observe that the numerical inf-sup/B-boundedness constant λ_B decreases with radius ε, that is, $\lambda_B = \lambda_B(\varepsilon)$.

As with the $2d$-$1d$ problem, we claim that λ_B is closely linked to a Steklov eigenvalue. In this case, we consider: Find $u \in V$, $\lambda_S > 0$ satisfying

$$\int_\Gamma \bar{u}\bar{v} = \lambda_S^2 \int_\Omega \nabla u \cdot \nabla v \quad \forall v \in V. \tag{26}$$

From (26) it follows that the maximal eigenvalue λ_S relates to the estimates of the mean value of u in Γ_ε, that is,

$$\|\bar{u}\|_Q \leq \lambda_S \|u\|_V \quad \forall u \in V.$$

The relation between the two eigenvalues is demonstrated in Fig. 17 which shows the relative error between the eigenvalues λ_B of the Schur complement of (25) and λ_S. Here, only the values obtained on the finest meshes for each ε are considered,

[6] The Brezzi conditions for the bilinear form a stemming from (25) are easily verified and the related constants take the value 1.

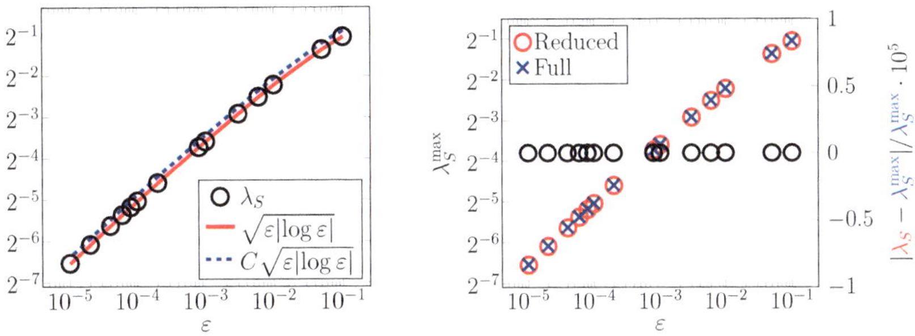

Fig. 18 (Left) Dependence of the Steklov eigenvalue $\lambda_{S,h_{\min}}$ (black $\circ$ markers) in (26) on the radius of coupling curve $\Gamma_\varepsilon = \{x \in \Omega, |x| = \epsilon\}$. The constant $C \approx 0.999$ was optimized for the best fit of the data. (Right) Comparison of the largest eigenvalues of the Steklov problems. The $2d$-$1d$ problem (16) is related to (23) with eigenvalues $\lambda_S^{\max} = \lambda_{S,h_{\min}}^{\max}$ (blue $\times$ markers), cf. Fig. 13, while λ_S (red $\circ$ markers) denotes eigenvalues of (26) related to $2d$-$0d$ problem (25), see Fig. 17. The relative error between the values of the full and reduced models is plotted against the right vertical axis in black markers $\circ$

that is, $\lambda_S = \lambda_{S,h_{\min}}$ and analogously for λ_B. In all cases, the observed error is $\sim 10^{-3}$.

Finally, in Fig. 18 we measure the dependence of λ_S (and λ_B) on the radius ε. It can be seen that the relation is practically identical to that of the unreduced $2d$-$1d$ problem (15), see also [25]. In particular, λ_B goes to 0 together with ε, and the affect of the inner radius on stability is not removed by model reduction.

In summary, we have established, through both numerical experiments and analytical expressions about the inf-sup and trace constant, a clear relationship between β and ε

$$\beta = C\sqrt{\varepsilon|\log\varepsilon|} \quad \text{as} \quad \varepsilon \to 0 \tag{27}$$

with C a constant independent from ε. The result appears independent from the discretization parameters of the considered numerical framework, so that we can assert with reasonable confidence that such behavior is inherent in the trace/extension operator structure of the interface problem.

5 Conclusion

The solvability of mixed-dimensional problems plays a crucial role in effectively applying these models to real-world scenarios, such as microcirculation. In our research, we have focused on operator preconditioning as a means to address this issue. We have shown that by employing suitable weighted norms, the operator preconditioning framework can successfully handle material parameters such as diffusivity in both the $3d$ and $1d$ domains. However, when dealing with interface-coupled systems, these norms alone are insufficient to ensure robustness regarding geometric parameters, such as the inner radius. Through extensive numerical experiments, we have highlighted the significant impact of the parameter ε on the mathematical structure of the problem and its adverse effect on the well-posedness through the trace operator. It is worth noting that this behavior persists even in a non-topologically reduced framework. Therefore, the reduction of dimensionality and the use of appropriately scaled Sobolev spaces currently do not guarantee the robustness of preconditioners as ε approaches zero. In our view, these findings strongly advocate a thorough and fundamental analysis of the trace operator's role in coupling conditions within mixed-dimensional approaches. The ultimate goal is to develop a generalized trace operator capable of facilitating a robust coupling of partial differential equations across high-dimensional gaps.

Acknowledgments Paolo Zunino acknowledges the support of the grant MUR PRIN 2022 No. 2022WKWZA8 *Immersed methods for multiscale and multiphysics problems* (IMMEDIATE). The present research is part of the activities of the Dipartimento di Eccellenza 2023–2027 project, funded by MUR. Nunzio Dimola and Paolo Zunino are members of the Gruppo Nazionale per il Calcolo Scientifico (GNCS), Istituto Nazionale di Alta Matematica (INdAM). Miroslav Kuchta gratefully acknowledges support from the Research Council of Norway, grant 303362.

Appendix

Numerical Experiments for Square-Shaped Inclusion

For the sake of comparison of the effect of the inclusion shape on the relation between the well-posedness of (16) and the diameter of the inclusion, we collect here additional results for $\Gamma_\varepsilon = \partial(-\varepsilon, \varepsilon)^2$ and $\Omega = (-1, 1)^2$. These results are analogous to Figs. 12 and 18 where circular Γ_ε is considered (Figs. 19 and 20).

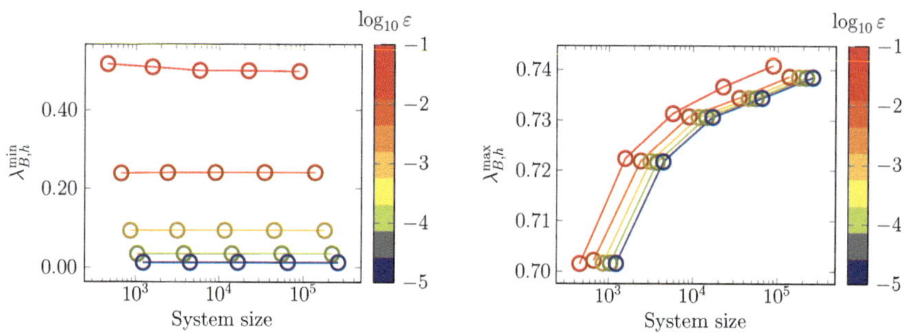

Fig. 19 (Left) Mesh convergence of the extremal eigenvalues of Schur complement (21) for $\Omega = (-1, 1)^2$, $\Gamma_\varepsilon = \partial(-\varepsilon, \varepsilon)^2$. Both spaces V and Q are discretized by $\mathbb{P}_1$ elements. For each ε a sequence of problems is considered on uniformly refined meshes starting from size $h_0 \geq h_l \geq h_{\min}$ leading to eigenvalues λ_{B,h_l}

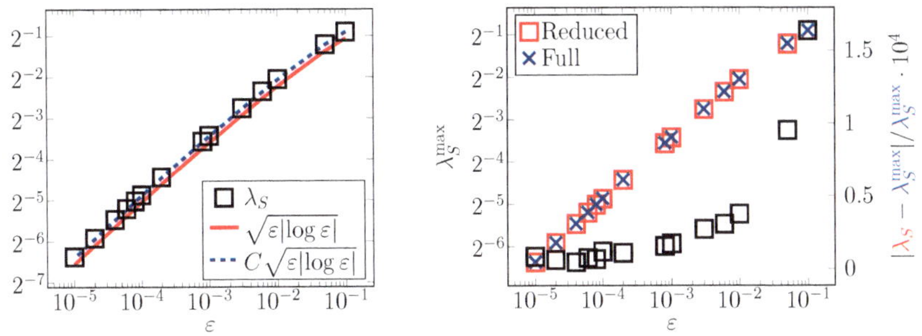

Fig. 20 Dependence of the Steklov eigenvalue $\lambda_{S,h_{\min}}$ in (26) on the radius of coupling curve $\Gamma_\varepsilon = \partial(-\varepsilon,\varepsilon)^2$. The constant $C \approx 1.116$ was optimized for the best fit of the data. (Right) Comparison of the largest eigenvalues of the Steklov problems. The $2d$-$1d$ problem (16) is related to (23) with eigenvalues $\lambda^{\max} = \lambda^{\max}_{S,h_{\min}}$, cf. Fig. 13, while λ_S denotes eigenvalues of (26) related to $2d$-$0d$ problem (25), see Fig. 17. The relative error between the values of the full and reduced models is plotted against the right vertical axis in black markers □

Numerical Experiments for Layered Mesh

Here we collect additional results regarding the numerical experiments for a circular domain $\Omega = \{x \in \mathbb{R}^2,\ |x| < 1\}$ with circular inclusion $\Gamma_\varepsilon = \{x \in \Omega,\ |x| = \varepsilon\}$ obtained on layered mesh (see Fig. 15 (right)) conformal simultaneously to every Γ_{ε_i}, with $\varepsilon_i \in \{10^{-1}, 10^{-2}, 10^{-3}, 10^{-4}, 10^{-5}\}$ (Figs. 21 and 22).

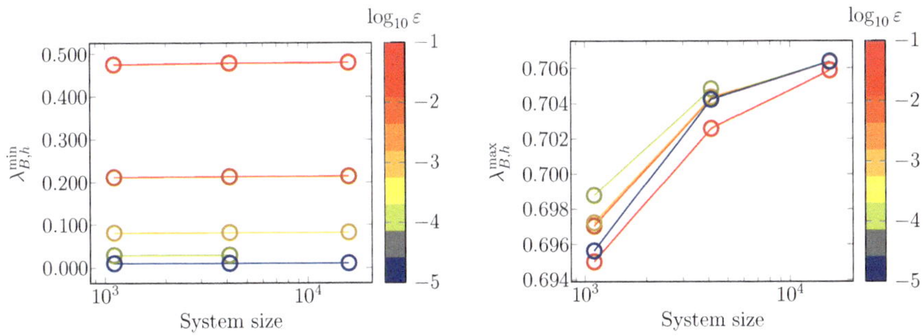

Fig. 21 (Left) Mesh convergence of the extremal eigenvalues of Schur complement (21) for $\Omega = \{x \in \mathbb{R}^2,\ |x| < 1\}$, $\Gamma_\varepsilon = \{x \in \Omega,\ |x| = \varepsilon\}$. Both spaces V and Q are discretized by $\mathbb{P}_1$ elements. For each ε a sequence of problems is considered on a uniformly refined layered mesh starting from size $h_0 \geq h_l \geq h_{\min}$ leading to eigenvalues λ_{B,h_l}

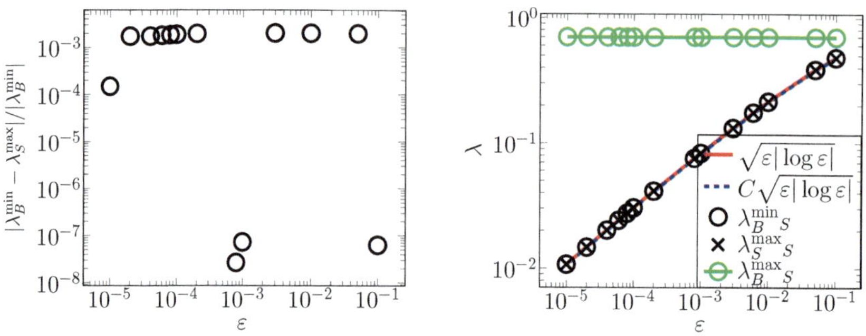

Fig. 22 Error between the smallest eigenvalue $\lambda_B^{\min}$ of the Schur complement problem (21) and the largest eigenvalue $\lambda_S^{\max}$ of the Steklov problem (23). In both cases, values of the finest level of refinement are considered, that is, $\lambda_X := \lambda_{X,h_{\min}}$. (Right) Dependence of the eigenvalues of the radius of the coupling curve $\Gamma_\varepsilon = \{x \in \Omega, |x| = \varepsilon\}$. The value $C \approx 0.999$ is obtained by fitting values $\lambda_B^{\min}$ for $\varepsilon < 10^{-1}$

Numerical Experiments for $\mathbb{P}_2 - \mathbb{P}_1$ Discretization

See Figs. 23 and 24.

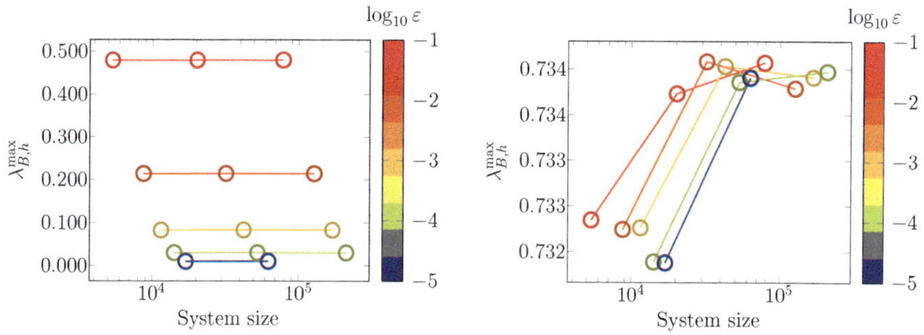

Fig. 23 (Left) Mesh convergence of the extremal eigenvalues of Schur complement (21) for $\Omega = \{x \in \mathbb{R}^2, |x| < 1\}$, $\Gamma_\varepsilon = \{x \in \Omega, |x| = \varepsilon\}$. Spaces V and Q are discretized respectively by $\mathbb{P}_2$ and $\mathbb{P}_1$ elements. For each ε a sequence of problems is considered on uniformly refined meshes starting from size $h_0 \geq h_l \geq h_{\min}$ leading to eigenvalues λ_{B,h_l}

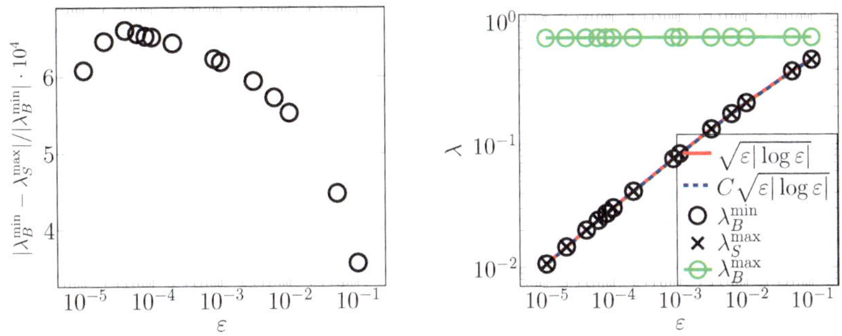

Fig. 24 Error between the smallest eigenvalue $\lambda_B^{\min}$ of the Schur complement problem (21) and the largest eigenvalue $\lambda_S^{\max}$ of the Steklov problem (23). In both cases, values of the finest level of refinement are considered, that is, $\lambda_X := \lambda_{X,h_{\min}}$. (Right) Dependence of the eigenvalues of the radius of the coupling curve $\Gamma_\varepsilon = \{x \in \Omega, |x| = \varepsilon\}$. The value $C \approx 0.999$ is obtained by fitting values $\lambda_B^{\min}$ for $\varepsilon < 10^{-1}$

References

1. Arnold, D.N., Rognes, M.E.: Stability of Lagrange elements for the mixed Laplacian. Calcolo **46**(4), 245–260 (2009)
2. Bærland, T., Kuchta, M., Mardal, K.A.: Multigrid methods for discrete fractional sobolev spaces. SIAM J. Sci. Comput. **41**(2), A948–A972 (2019). https://doi.org/10.1137/18M1191488
3. Blake, T., Gross, J.: Analysis of coupled intra- and extraluminal flows for single and multiple capillaries. Math. Biosci. **59**(2), 173–206 (1982)
4. Braess, D.: Stability of saddle point problems with penalty. Math. Model. Numer. Anal. Modél. Math. Anal. Numér. **30**(6), 731–742 (1996)
5. Bramble, J.H., Pasciak, J.E., Vassilevski, P.S.: Computational scales of Sobolev norms with application to preconditioning. Math. Comput. **69**(230), 463–480 (2000). https://doi.org/10.1090/S0025-5718-99-01106-0
6. Brezzi, F.: On the existence, uniqueness and approximation of saddle-point problems arising from Lagrangian multipliers. Publ. Math. Inf. Rennes **8**, 129–151 (1974)
7. Budiša, A., Hu, X., Kuchta, M., Mardal, K.A., Zikatanov, L.: Rational approximation preconditioners for multiphysics problems. In: International Conference on Numerical Methods and Applications, pp. 100–113. Springer, Berlin (2022)
8. Budiša, A., Hu, X., Kuchta, M., Mardal, K.A., Zikatanov, L.: Rational approximation preconditioners for multiphysics problems. In: Georgiev, I., Datcheva, M., Georgiev, K., Nikolov, G. (eds.) Numerical Methods and Applications, pp. 100–113. Springer, Cham (2023)
9. Cerroni, D., Laurino, F., Zunino, P.: Mathematical analysis, finite element approximation and numerical solvers for the interaction of 3d reservoirs with 1d wells. GEM-Int. J. Geomath. **10**, 1–27 (2019)
10. D'Angelo, C.: Multi scale modelling of metabolism and transport phenomena in living tissues, PhD Thesis. EPFL, Lausanne (2007)
11. D'Angelo, C.: Finite element approximation of elliptic problems with Dirac measure terms in weighted spaces: applications to one-and three-dimensional coupled problems. SIAM J. Numer. Anal. **50**(1), 194–215 (2012)
12. D'Angelo, C., Quarteroni, A.: On the coupling of 1d and 3d diffusion-reaction equations: application to tissue perfusion problems. Math. Models Methods Appl. Sci. **18**(08), 1481–1504 (2008)

13. Fleischman, G., Secomb, T., Gross, J.: The interaction of extravascular pressure fields and fluid exchange in capillary networks. Math. Biosci. **82**(2), 141–151 (1986)
14. Flieschman, G., Secomb, T., Gross, J.: Effect of extravascular pressure gradients on capillary fluid exchange. Math. Biosci. **81**(2), 145–164 (1986)
15. Formaggia, L., Vergara, C.: Defective Boundary Conditions for PDEs with Applications in Haemodynamics, pp. 285–312. Springer, Cham (2018)
16. Führer, T.: Multilevel decompositions and norms for negative order Sobolev spaces. Math. Comput. **91**(333), 183–218 (2021). https://doi.org/10.1090/mcom/3674
17. Gjerde, I.G., Kumar, K., Nordbotten, J.M.: A singularity removal method for coupled 1D–3D flow models. Comput. Geosci. **24**, 443–457 (2020)
18. Gregorio, S.D., Possenti, L., di Gregorio, S., Gerosa, F.M., Raimondi, G., Casagrande, G., Costantino, M.L., Zunino, P.: A computational model for microcirculation including fahraeus-lindqvist effect, plasma skimming and fluid exchange with the tissue interstitium. Int. J. Numer. Methods Biomed. Eng. **35**(3), e3165 (2019)
19. Hartung, G., Badr, S., Moeini, M., Lesage, F., Kleinfeld, D., Alaraj, A., Linninger, A.: Voxelized simulation of cerebral oxygen perfusion elucidates hypoxia in aged mouse cortex. PLOS Comput. Biol. **17**(1), 1–28 (2021). https://doi.org/10.1371/journal.pcbi.1008584
20. Hersch, J., Payne, L.E., Schiffer, M.M.: Some inequalities for Stekloff eigenvalues. Arch. Rational Mech. Anal. **57**(2), 99–114 (1974)
21. Hu, X., Keilegavlen, E., Nordbotten, J.M.: Effective preconditioners for mixed-dimensional scalar elliptic problems. Water Resources Res. **59**(1), e2022WR032985 (2023)
22. Koch, T., Heck, K., Schröder, N., Class, H., Helmig, R.: A new simulation framework for soil–root interaction, evaporation, root growth, and solute transport. Vadose Zone J. **17**(1), 1–21 (2018)
23. Koch, T., Schneider, M., Helmig, R., Jenny, P.: Modeling tissue perfusion in terms of 1d-3d embedded mixed-dimension coupled problems with distributed sources. J. Comput. Phys. **410**, 109370 (2020)
24. Köppl, T., Wohlmuth, B.: Optimal a priori error estimates for an elliptic problem with Dirac right-hand side. SIAM J. Numer. Anal. **52**(4), 1753–1769 (2014)
25. Köppl, T., Vidotto, E., Wohlmuth, B., Zunino, P.: Mathematical modeling, analysis and numerical approximation of second-order elliptic problems with inclusions. Math. Models Methods Appl. Sci. **28**(5), 953–978 (2018)
26. Kuchta, M.: Assembly of multiscale linear PDE operators. In: Numerical Mathematics and Advanced Applications ENUMATH 2019, pp. 641–650. Springer, Berlin (2021)
27. Kuchta, M., Nordaas, M., Verschaeve, J., Mortensen, M., Mardal, K.: Preconditioners for saddle point systems with trace constraints coupling 2d and 1d domains. SIAM J. Sci. Comput. **38**(6), B962–B987 (2016)
28. Kuchta, M., Mardal, K.A., Mortensen, M.: Preconditioning trace coupled 3d-1d systems using fractional Laplacian. Numer. Methods Partial Differ. Equ **35**, 375–393 (2019)
29. Kuchta, M., Laurino, F., Mardal, K.A., Zunino, P.: Analysis and approximation of mixed-dimensional PDEs on 3D-1D domains coupled with Lagrange multipliers. SIAM J. Numer. Anal. **59**(1), 558–582 (2021)
30. Laurino, F., Zunino, P.: Derivation and analysis of coupled PDEs on manifolds with high dimensionality gap arising from topological model reduction. ESAIM: M2AN **53**(6), 2047–2080 (2019)
31. Lions, J.L., Magenes, E.: Non-homogeneous Boundary Value Problems and Applications: Vol. 1, vol. 181. Springer, Berlin (2012)
32. Logg, A., Mardal, K.A., Wells, G.N., et al.: Automated Solution of Differential Equations by the Finite Element Method. Springer, Berlin (2012)
33. Mardal, K.A., Winther, R.: Preconditioning discretizations of systems of partial differential equations. Numer. Linear Algebra Appl. **18**(1), 1–40 (2011)
34. Masri, R., Zeinhofer, M., Kuchta, M., Rognes, M.E.: The modelling error in multi-dimensional time-dependent solute transport models (2023). arXiv:2303.17999

35. Maury, B.: Numerical analysis of a finite element/volume penalty method. SIAM J. Numer. Anal. **47**(2), 1126–1148 (2009). https://doi.org/10.1137/080712799
36. Murphy, M.F., Golub, G.H., Wathen, A.J.: A note on preconditioning for indefinite linear systems. SIAM J. Sci. Comput. **21**(6), 1969–1972 (2000)
37. Nečas, J.: Direct Methods in the Theory of Elliptic Equations. Springer, Berlin (2011)
38. Nitsche, J.: Über ein variationsprinzip zur lösung von dirichlet-problemen bei verwendung von teilräumen, die keinen randbedingungen unterworfen sind. Abhandlungen aus dem Mathematischen Seminar der Universität Hamburg **36**(1), 9–15 (1971). https://doi.org/10.1007/BF02995904
39. Peaceman, D.: Interpretation of well-block pressures in numerical reservoir simulation. Soc. Pet. Eng. AIME J. **18**(3), 183–194 (1978)
40. Peaceman, D.W.: Interpretation of well-block pressures in numerical reservoir simulation with nonsquare grid blocks and anisotropic permeability. Soc. Petroleum Eng. J. **23**(3), 531–543 (1983)
41. Qin, J.: On the convergence of some low order mixed finite elements for incompressible fluids. The Pennsylvania State University (1994)
42. Rognes, M.E.: Automated testing of saddle point stability conditions. In: Automated Solution of Differential Equations by the Finite Element Method: The FEniCS Book, pp. 657–671. Springer, Berlin (2012)
43. Rusten, T., Winther, R.: A preconditioned iterative method for saddlepoint problems. SIAM J. Matrix Anal. Appl. **13**(3), 887–904 (1992)
44. Saad, Y.: Iterative Methods for Sparse Linear Systems. SIAM, Philadelphia (2003)
45. Secomb, T., Hsu, R., Park, E., Dewhirst, M.: Green's function methods for analysis of oxygen delivery to tissue by microvascular networks. Ann. Biomed. Eng. **32**(11), 1519–1529 (2004)
46. Steinbach, O.: Numerical Approximation Methods for Elliptic Boundary Value Problems: Finite and Boundary Elements. Springer, Berlin (2007)
47. Ventimiglia, T., Linninger, A.A.: Mesh-free high-resolution simulation of cerebrocortical oxygen supply with fast fourier preconditioning. Int. J. Numer. Methods Biomed. Eng. **39**(8), e3735 (2023). https://doi.org/10.1002/cnm.3735

Numerical Approaches for Multiphase Microfluids

Alessandro Coclite and Francesco De Vita

Abstract In this contribution strength and limitations of two modern and reliable numerical models for computing multiphase flows in the microcirculation regimens are reviewed and critically analyzed. Specifically, *algebraic Volume of Fluid* approach for fluid/fluid interaction problems is considered, as well as the *dynamic-Immersed Boundary* for fluid/structure interaction problems with a special accent on bio-inspired systems. On one hand, the one-fluid approach for computing multi species mixtures is adopted. The Navier-Stokes system for the mixture evolution is considered and the interface between the different species is tracked through a convenient discretization of the Heaviside function. On the other hand, the dynamic Immersed Boundary method combined with a BGK Lattice–Boltzmann technique is presented, in which the body surface is modeled as a collection of points responding to an elastic potential and a bending resistance. A moving least squares reconstruction is used to accurately interpolate flow quantities and the forcing field needed to enforce the boundary condition on immersed bodies. Validations against well known benchmark data for both models are discussed.

Keywords Dynamic immersed boundary · Microfluidics · Volume of fluid · Multiphase flows · Numerical methods

1 Introduction

Microfluidic is the science of fluids at *small* scales. Small, here, is typically used for flows in which at least one of the length scales ranges between micrometers and nanometers. This implies that the dominant physics in microfluidic differs from

A. Coclite (✉)
Dipartimento di Ingegneria Elettrica e dell'Informazione (DEI), Politecnico di Bari, Bari, Italy
e-mail: alessandro.coclite@poliba.it

F. De Vita
Dipartimento di Meccanica, Matematica e Management (DMMM), Politecnico di Bari, Bari, Italy
e-mail: francesco.devita@poliba.it

A. Linninger et al. (eds.), *Quantitative Approaches to Microcirculation*,
SEMA SIMAI Springer Series 36, https://doi.org/10.1007/978-3-031-58519-7_7

that for large scale flows. For example, microfluidic flows are in general laminar, hence viscous forces are comparable to or more important than inertial forces; other forces, such as surface tension, which are typical negligible at large scales can play an important role at small scales, affecting significantly the flow dynamics. Nevertheless, microfluidic flows are governed by the classical fluid mechanics equations and numerical simulations are an important tool to study this flows.

Historically, microfluidics originate from four different fields: molecular analysis, biodefence, molecular biology and microelectronics [1]. Starting from those fields it has spread and nowadays has several applications such as the generation control of bubbles or droplet [2, 3], mixing enhancement [4], gene sequencing and Point-of-Care diagnostics [5–10], Lab-on-chip devices development [11], cell separation and sorting for diagnostic and therapeutic applications [12]. In many of the aforementioned applications the system involves two or more phases (solid, liquid or gas), so that, the fluid evolving in the microfluidic device is assumed as a dispersion. A dispersion is a multiphase system composed of small particles, droplets or bubbles dispersed in a matrix phase. Typically, the size of the dispersed phase ranges between nanometers and micrometers [13]. Depending on the involved phases, such fluids are classified as: *emulsion*, a liquid dispersion in a liquid (for example mayonnaise or skin creams); *foam*, a gas dispersion in a liquid (for example champagne or shampoo); *suspension*, a solid dispersion in a liquid (for example paints, mud or polymer solutions); *aerosol*, a solid or liquid dispersion in a gas (for example ink jets or sprays). Lyophobic colloids are thermodynamically unstable and in this case the interaction of the dispersed phase can induce some changes in the material properties. Dispersion transport can be due to flow advection, buoyancy sedimentation or Brownian motion. Dispersions are usually referenced as *non-colloidal* when the latter is negligible and *colloidal* in the opposite case. The two regimens discriminated by the Peclet number, the ratio between advection and diffusion forces. For example, in a sheared suspension of neutrally buoyant spherical particles with radius a, thermal diffusion is proportional to $k_B T/a$, T being the temperature of the suspension and k_B the Boltzmann constant; advection is given by the shear rate $\dot{\gamma}$ and the hydrodynamic force is proportional to $6\pi\mu\dot{\gamma}a^2$, where μ is the viscosity of the liquid phase. The resulting Peclet number is $Pe = 6\pi\mu\dot{\gamma}a^3/k_B T$ and the dispersion can be typically considered non-colloidal for particle sizes larger than $1\mu m$ and shear rate larger than $\dot{\gamma} = 0.1\ s^{-1}$ [14]. For emulsions under buoyancy the Peclet number can be expressed as $Pe = 4\pi\mu a^4 \Delta\rho g/3k_B T$, with $\Delta\rho$ the density difference between the two liquid phases and g the gravity acceleration, showing that Brownian motion can be negligible for large droplet sizes and large density ratios [15]. This study will focus on non-colloidal dispersions.

Macroscopic properties of a dispersion depends on its microscopic properties; such as, dispersed phase total exposed surface and size distribution, porosity, presence of defects, internal stresses and state of agglomeration. Nevertheless, on a large enough scale the dispersion can be considered as a continuous fluid with effective material properties which in general behaves as a non-Newtonian fluid. One of the most important effect of the presence of colloids in a fluid matrix is a change in the viscosity of the mixture, referred as the *effective viscosity*. This has relevant consequences for applications since dissipation depends on viscosity and an increased effective viscosity leads to high energy consumption, for example, among many other fields, in fabrication of cosmetics, food processing and oil extractions.

The study of the effective viscosity of a suspension dates back to the pioneering work of Einstein [16] who proposed to model the effective viscosity of a dilute mono-dispersion of rigid spheres as $\mu_e = \mu_m (1 + \phi)$, with μ_m the viscosity of the matrix phase and ϕ the volume fraction of the spheres (i.e the volume of the spheres over the volume of the dispersion). This model was accurately enriched by Einstein [17] giving the expression $\mu_e = \mu_m (1 + 5\phi/2)$. The increases in the effective viscosity is due to the flow disturbance induced by the rigid spheres resisting shear deformation, which correspond to an increased viscous dissipation. At vanishing inertia (small Reynolds number), shear stress is proportional to shear rate and the suspension can be seen as a Newtonian fluid with effective viscosity as a function only of the volume fraction $\mu_e = \mu_e (\phi)$. This expression returns a good approximation of the effective viscosity for an extremely dilute suspension (volume fraction up to 0.05); for larger values of the volume fraction, pair interactions need to be accounted. Batchelor and Green [18], assuming a random distribution of particles, found a correction in the effective viscosity of the order $O(\phi^2)$ giving $\mu_e = \mu_m \left(1 + 5\phi/2 + 5\phi^2\right)$. This formula reasonably agrees with experimental data up to $\phi \approx 0.15$ while, for larger values of the volume fraction no closed expression are available and experimental fit must be used; an accurate expression was proposed by Eilers [19] as $\mu_e = \mu_m [1 + (5\phi/4) / (1 - \phi/\phi_c)]$, with ϕ_c representing the maximum packing volume fraction. For higher volume fractions, the suspension behaves as non-Newtonian: shear thickening, for example, has been observed by Mari et al. [20] who studied suspensions in the contact dominated regime. A detailed review of the rheology of dense suspension of rigid particles can be found in [21]. Moreover, A. Pries et al. proposed an explicit expression for the effective viscosity as a function of the vessel radius and hematocrit level by correlating the mechanical behavior of blood borne cells with the global properties of the blood due to its specific constituents [22].

All previous relations assume particles to be rigid. Taylor [23] firstly investigated the influence of deformability on the effective viscosity: he considered a dilute dispersion of droplets of one fluid in a matrix phase of a different fluid and introduced the ratio between dispersed and matrix phase viscosity λ into the Einstein's formula as follow $\mu_e = \mu_m [1 + 2.5\phi (\lambda + 0.4) / (\lambda + 1)]$. Pal [24] proposed expressions for the two limiting regimens of infinitely dilute and concentrated emulsions based on the effective medium approach. Recently, numerical simulations provided results for the effective viscosity and normal stresses in the moderate and high viscosity ratio regime [25] as well as in the dense regime [26] investigating the effect of the capillary number. Rosti et al. [27] proposed a scaling of the effective viscosity considering the deformability as an effective volume fraction effect. Maffettone et al. [28, 29] systematically reviewed the role of hydrodynamic interactions in terms of single-particle problems, binary interactions, and multi-body systems in the dynamics of rigid particles transported in a viscoelastic fluids. Abtahi et al. [30] investigated the dynamics of an isolated ellipsoid in a sheared Carreau fluid thus demonstrating the degeneration of the Jeffery orbits when considering shear-thinning rheology in the same fashion of what observed in Newtonian fluids. When studying dispersion of capsule-like particles, deformability is the main physical mechanism to consider. On the other sides, when the dispersed phase consist of droplets (as for emulsions), also coalescence and breakup play an important role on the effective viscosity. In a dilute sheared emulsion, the dynamic of droplets is dictated by the equilibrium between the viscous stress, generated by the external flow, and the interfacial tension. The ratio between this two forces is represented by the capillary number (Ca). When the Ca of a droplet exceed a critical value, it breaks in order to reach again balance between the two forces. In the non-dilute regime, collisions occur between droplets and the outcome of a collision depends on different aspect. Chesters [31] proposed the following scenario: when the collision time is larger than the drainage time, the collision ends up into droplet coalescence; in the opposite case droplets repel. The behavior of droplets under shear flow has been extensively studied [32, 33]; recently, Casterta and Guido [34], experimentally, and De Vita et al. [35], numerically, have demonstrated that emulsions can exhibit a negative curvature of the rheological curve effective viscosity vs volume fraction. This effect is consequence of coalescence which reduces the total surface of the dispersed phase, hence, there is a reduction in the total surface energy. This behavior is in opposition to suspensions, where the curvature of the rheological curve μ_e vs ϕ display a positive curvature.

On this basis, clearly emerges the pressing need for numerical tools and computational models able to accurately approximate the dynamics of the dispersed phase immersed in the solvent phase. Several numerical methods can be adopted to model these physical properties; in the following we will focus on the Volume of Fluid (VoF) method by Rosti et al. [36] and the dynamic-Immersed Boundary (dynamic-IB) framework by Coclite et al. [37]. Strength and limitations of both approaches will be critically discussed. The present paper is organized as follows. In Sect. 1 an historical perspective on the modeling of the effective viscosity is given. Section 2 discuss the algebraic Volume of Fluid approach for analyzing the physical features of multiphase fluids (fluid/fluid interaction). Section 3 describe a well consolidate approach for fluid/structure interaction, the dynamic-Immersed Boundary approach, along with a critical discussion on its strength and limitations. Conclusive remarks and future outlooks end the discussion.

2 Fluid/Fluid Interaction

Multiphase fluid-fluid systems can be simulated by solving Navier-Stokes equation in each fluid domain $\Omega_i \subset R^3$, with $i = 1, \ldots, N$ representing the *i-th* fluid phase, and applying proper boundary condition at the boundary $\partial\Omega_i$, which can be a challenging computational problem. To overcome this complexity, a commonly used approach is the "one-fluid" formulation: in this method the two fluids system is considered as one single fluid evolving in $\Omega = \bigcup_{i=1}^{N} \Omega_i$, with material properties varying across the interface separating the phases $\partial\Omega_i$. The location of the interface $\partial\Omega_i$ can be identified by introducing an Heaviside function $\mathcal{H} = \mathcal{H}(\mathbf{x}, t)$ so that $\mathcal{H} = 0$ in one fluid and $\mathcal{H} = 1$ in the other. Material properties of the "one-fluid", then, are linked to those of the phases as

$$\begin{aligned} \rho\left(\mathcal{H}(\mathbf{x}, t)\right) &= \rho_1 \mathcal{H}(\mathbf{x}, t) + (1 - \mathcal{H}(\mathbf{x}, t))\rho_2 \\ \mu\left(\mathcal{H}(\mathbf{x}, t)\right) &= \mu_1 \mathcal{H}(\mathbf{x}, t) + (1 - \mathcal{H}(\mathbf{x}, t))\mu_2 \end{aligned} \tag{1}$$

with $\rho_{1,2}$ and $\mu_{1,2}$ the density and viscosity of the two fluid phases. Numerically, the Heaviside function is replaced by its cell average value as

$$\mathcal{F}(\mathbf{x}, t) = \frac{1}{\delta\mathcal{V}} \int_{\delta\mathcal{V}'} \mathcal{H}\left(\mathbf{x}', t\right) \delta\mathcal{V}' \tag{2}$$

and represent the volume fraction, or Volume of Fluid function, of a fluid phase in a control volume $\delta\mathcal{V}$; cells where $0 < \mathcal{F} < 1$ are cut by the interface. In practice, this condition is replaced with $\varepsilon < \mathcal{F} < 1 - \varepsilon$ with ε a given numerical tolerance (typical it is assumed $\varepsilon = 10^{-8}$). For constant density fluid flows, and in absence of phase changes, the velocity vector is continuous across the interface and the Volume of Fluid function $\mathcal{F}$ is simply advected by the flow as

$$\partial_t \mathcal{F}(\mathbf{x}, t) + \nabla \cdot \left(\mathbf{u}(\mathbf{x}, t) \mathcal{H}(\mathbf{x}, t)\right) = \mathcal{F}(\mathbf{x}, t) \nabla \cdot \mathbf{u}(\mathbf{x}, t), \quad \mathbf{x} \in \Omega, t > 0 \tag{3}$$

with $\mathbf{u}(\mathbf{x}, t)$ the velocity vector field. Finally, the dynamic of the "one-fluid" is computed by solving the continuity equation

$$\nabla \cdot \mathbf{u}(\mathbf{x}, t) = 0, \quad \mathbf{x} \in \Omega, t > 0 \tag{4}$$

and the momentum balance equation

$$\begin{aligned} \rho\left(\mathcal{H}(\mathbf{x}, t)\right)\left[\partial_t \mathbf{u}(\mathbf{x}, t) + (\mathbf{u}(\mathbf{x}, t) \cdot \nabla)\, \mathbf{u}(\mathbf{x}, t)\right] &= -\nabla p(\mathbf{x}, t) + \\ +\rho \mathbf{f}(\mathbf{x}, t) + \nabla \cdot \left(2\mu\left(\mathcal{H}(\mathbf{x}, t)\right) \mathbf{D}(\mathbf{x}, t)\right) &+ \sigma\kappa\delta_S(\mathbf{x}, t)\, \mathbf{n}, \quad \mathbf{x} \in \Omega, t > 0 \end{aligned} \tag{5}$$

with $p = p(\mathbf{x}, t)$ the pressure field, $\mathbf{D}(\mathbf{x}, t)$ the strain rate tensor defined as $D_{ij}(\mathbf{x}, t) = \left(\partial_j u_i(\mathbf{x}, t) + \partial_i u_j(\mathbf{x}, t)\right)/2$, $\mathbf{f}(\mathbf{x}, t)$ the vector field of volume forces (*e.g.* gravity). The jump condition due to surface tension at $\partial\Omega_i$ is accounted by the last term at *r.h.s* of Eq. (5) where σ is the surface tension coefficient, κ and $\mathbf{n}$ the local curvature and normal to the interface and the δ_S is the Dirac distribution, expressing that the surface tension forces is applied only across the interface. Note that, given $\mathcal{H}$ as an Heaviside function, the delta Dirac function is given by $\delta_S \mathbf{n} = \nabla\mathcal{H}$. The key aspect of the Volume of Fluid method is to accurately solve the advection equation (3). Different Volume of Fluid methods have been developed in the past, based on the approximation of the Heaviside function $\mathcal{H}$: geometrical VoF methods use a piecewise-linear geometrical reconstruction of the interface [38, 39], while algebraic methods relay on the approximation of the Heaviside function via an hyperbolic tangent function [40, 41]. Below, the latter approach will be described.

2.1 Tangent of Hyperbola Interface Capturing (THINC) Method

In the algebraic Volume of Fluid method [41], the Heaviside function is approximated as

$$\mathcal{H}(\mathbf{x},t) \approx \hat{\mathcal{H}}(\mathbf{x},t) = \frac{1}{2}\left(1+\tanh\left(b\left(P\left(\mathbf{x},t\right)+d\right)\right)\right) \tag{6}$$

where b is a sharpness parameter, $P(\mathbf{x})$ a surface function (either linear or quadratic) and d a shift parameter used to enforce mass conservation, i.e.

$$\int_{\partial\mathcal{V}'} \hat{\mathcal{H}}(\mathbf{x},t)\,\partial\mathcal{V}' = \mathcal{F}(\mathbf{x},t)\,. \tag{7}$$

The domain Ω is approximated by a Cartesian grid with grid spacing Δx, Δy, Δz in each direction and time is discretized with time step Δt. Hence, $\mathcal{F}(\mathbf{x},t) \equiv \mathcal{F}((i\Delta x, j\Delta y, k\Delta z), n\Delta t) \equiv \mathcal{F}^n_{i,j,k}$. Given the solution at time step n, Eq. (3) is discretized using the directional splitting as follow using the directional splitting:

$$\mathcal{F}^*_{ijk} = \mathcal{F}^n_{ijk} - \frac{1}{\Delta x}\left(a^n_{i+1/2jk} - a^n_{i-1/2jk}\right) + \frac{\Delta t}{\Delta x}\mathcal{F}^*_{ijk}\left(u^n_{i+1/2jk} - u^n_{i-1/2jk}\right) \tag{8a}$$

$$\mathcal{F}^{**}_{ijk} = \mathcal{F}^*_{ijk} - \frac{1}{\Delta y}\left(b^*_{ij+1/2k} - b^*_{ij-1/2k}\right) + \frac{\Delta t}{\Delta y}\mathcal{F}^{**}_{ijk}\left(v^n_{ij+1/2k} - v^n_{ij-1/2k}\right) \tag{8b}$$

$$\mathcal{F}^{***}_{ijk} = \mathcal{F}^{**}_{ijk} - \frac{1}{\Delta z}\left(c^{**}_{ijk+1/2} - c^{**}_{ijk-1/2}\right) + \frac{\Delta t}{\Delta z}\mathcal{F}^{***}_{ijk}\left(w^n_{ijk+1/2} - w^n_{ijk-1/2}\right) \tag{8c}$$

$$\begin{aligned}\mathcal{F}^{n+1}_{ijk} &= \mathcal{F}^{***}_{ijk} - \Delta t\,(\,\mathcal{F}^*_{ijk}\frac{\left(u^n_{i+1/2jk} - u^n_{i-1/2jk}\right)}{\Delta x} + \\ &+\mathcal{F}^{**}_{ijk}\frac{\left(v^n_{ij+1/2k} - v^n_{ij-1/2k}\right)}{\Delta y} + \mathcal{F}^{***}_{ijk}\frac{\left(w^n_{ijk+1/2} - w^n_{ijk-1/2}\right)}{\Delta z}\,)\end{aligned} \tag{8d}$$

where the flux terms a^n, b^* and c^{**} are given by

$$a^n_{i\pm1/2jk} = \frac{1}{\Delta y \Delta z} \int_{\delta t^n} \int_{\delta y} \int_{\delta z} (u(\mathbf{x}, t)\mathcal{H}(\mathbf{x}, t))_{i\pm1/2jk}\, dydzdt \tag{9a}$$

$$b^*_{ij\pm1/2k} = \frac{1}{\Delta x \Delta z} \int_{\delta t^*} \int_{\delta x} \int_{\delta z} (v(\mathbf{x}, t)\mathcal{H}(\mathbf{x}, t))_{ij\pm1/2k}\, dxdzdt \tag{9b}$$

$$c^{**}_{ijk\pm1/2} = \frac{1}{\Delta x \Delta y} \int_{\delta t^{**}} \int_{\delta x} \int_{\delta y} (w(\mathbf{x}, t)\mathcal{H}(\mathbf{x}, t))_{ijk\pm1/2}\, dxdydt \tag{9c}$$

with $\delta t^n = \delta t^* = \delta t^{**} = \Delta t$. In practice, time integration is replaced with spatial integration in the upwind direction and the integration is performed analytically in the direction of the normal and via Gaussian quadrature in the other two directions [41]. The choice of the value of the sharpness parameter b can affect the interface evolution: a small value of the parameter leads to volume of fluid function smearing, whereas a large value can lead to interface wiggles. Ii et al. [41] have studied the influence of the sharpness parameter on the numerical error for the interface evolution; a detailed discussion is also reported in [42].

2.2 *Pressure Poisson Equation*

For single phase incompressible fluid flows, continuity equation (4) does not contain time derivatives hence the mass conservation constrain must be obtained by an implicit coupling via the pressure in the momentum equation [43]. This is usually obtained first computing a provisional velocity field $\mathbf{u}^*$ from momentum equation as

$$\mathbf{u}^*_{ijk} = \mathbf{u}^n_{ijk} - \frac{\Delta t}{\rho^{n+1/2}_{ijk}} [\,\nabla p^n_{ijk} + (-\mathbf{u}_{ijk} \cdot \nabla \mathbf{u}_{ijk} + \nabla \cdot \left(2\mu_{ijk}\mathbf{D}_{ijk}\right))^{n+1/2}\,] \tag{10}$$

and then projecting $\mathbf{u}^*$ into a divergence free velocity field as

$$\mathbf{u}^{n+1} = \mathbf{u}^n - \frac{\Delta t}{\rho}\nabla\phi \tag{11}$$

with ϕ a scalar function representing the pressure increment in the time step $p^{n+1} = p^n + \phi$. By construction, the scalar function ϕ must obey the Poisson equation

$$\nabla \cdot \left(\frac{\Delta t}{\rho}\nabla\phi\right) = \nabla \cdot \mathbf{u}^* \tag{12}$$

which for single phase incompressible flows is a constant coefficient Poisson equation and Fast Direct Solver can be used to get accurate (machine precision) divergence free velocity field. When solving for multiphase flows, the density is non-constant but it is function of the interface location $\mathcal{H}$. In this case, by introducing the pressure splitting [44] one can still get a constant coefficient approximated Poisson equation and employ a Fast Direct Solver. See [36, 45] for a detailed description of the method and validations.

2.3 Applications in Microfluidics

As mentioned in the introduction, microfluidic applications are usually dominated by viscous and surface tension forces. Hence an accurate description of interface normal and curvature is needed to properly evaluate surface tension effects. First we study the shape evolution of a single droplet in a shear flow. The droplet, with initial circular shape and radius a, is located at the center of a channel with top and bottom walls moving with opposite velocity $\pm U$. In Fig. 1 we report the time evolution of the deformation parameter $D = (L - B)/(L + B)$, with L and B major and minor axis of the ellipsoid, for three values of the capillary number $Ca = U\mu/\sigma$, being μ the fluid viscosity and σ the surface tension coefficient. Both viscosity ratio and Reynolds number $Re = \rho U 2a/\mu$ are unitary. We compare results obtained with the THINC solver against data available in literature [46] and with data obtained with the open source solver *basilisk* (http://basilisk.fr/) which is based on a geometrical VoF [39]. Data obtained with the THINC solver and *basilisk* are in excellent agreement, while the small difference with respect to the literature data is consequence of the larger spreading of surface force adopted in [46].

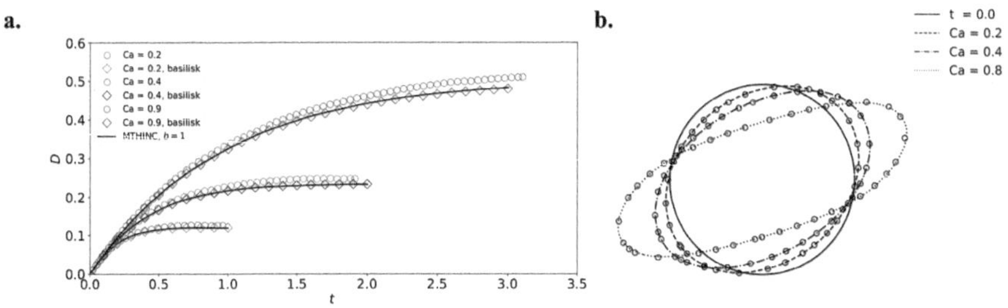

Fig. 1 Deformation of a drop under shear. (**a**) Time history of the deformation parameter D for different values of the Capillary number: (empty circle) data from [46]; (empty diamond) data obtained with *basilisk*; (solid line) data obtained with the THINC method. (**b**) Steady state droplet shape comparison between THINC method (lines) and *basilisk* (empty circle)

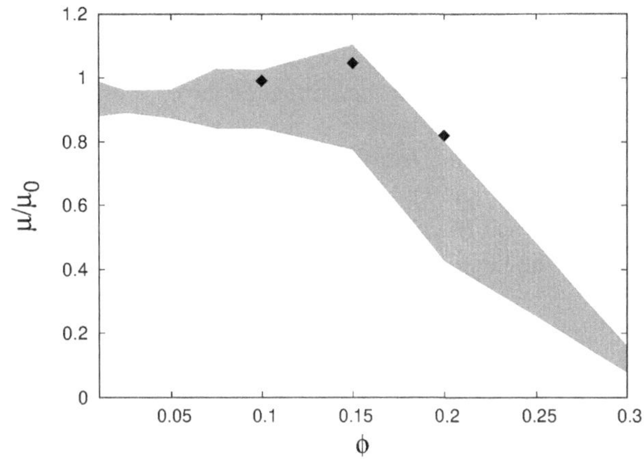

Fig. 2 Emulsion effective viscosity with respect to volume fraction ϕ. Shaded area corresponds to the cloud of available experimental data; symbols are simulation results

2.4 Emulsions Effective Viscosity

The THINC solver has been recently applied to study rheology of emulsions under shear flow [35] and to explain the experimentally observed shear bending behavior [47]. Caserta and Guido [34] have reported that, for a dispersed phase viscosity lower than the carrier fluid viscosity, a sheared emulsion can exhibit shear bending. This effect have been demonstrated to be consequence of coalescence: the reduction of the total surface lead to a reduction of surface energy with consequence reduction of effective viscosity [47]. This demonstrate that coalescence must be accounted when evaluating effective viscosity of a dispersion, and all available relation do not properly reproduce this effect. Figure 2 shows the curve effective viscosity vs volume fraction for an emulsion with dispersed phase viscosity lower than carrier fluid viscosity (μ_0). The black diamonds represent the value of the effective viscosity computed with numerical simulations, and fall within the cloud of experimental values (gray are).

3 Fluid/Structure Interaction: The dynamic-IB Approach

In this section the numerical procedure designed by Coclite et al. [37, 48, 49] for studying the evolution of a structure immersed into a viscous flow is detailed. A finite difference framework is developed relying on the strong coupling between the fluid and the structure dynamics. The two solutions are obtained on two separate meshes interacting via a transfer function for entrusting, on the fluid side, no-slip conditions over the immersed structure; while, on the structure side, external hydrodynamics stresses. Specifically, the evolution of fluid conserved quantities is obtained through the solution of the Boltzmann equation in the BGK approximation on a 3D Cartesian lattice endowed with 19 discrete velocity directions. The Boltzmann equation is enriched with a forcing term demanded for the fluid/structure coupling. Zou-He conservative bounce-back procedure [50] is used for enforcing Dirichlet known-velocity conditions on external boundaries while a dynamic-Immersed Boundary approach defines no-slip conditions over the immersed structure. The latter being integrated in time with a P-(EC)^k β-Newmark scheme.

Let $\Omega \subset \mathbb{R}^3$ be a tridimensional (3D) fluid domain physically characterized by the density $\rho_f > 0$ and the kinematic viscosity $\nu > 0$. For every $t \geq 0$ (t being the time coordinate), the fluid velocity field is given by $\mathbf{u} : \Omega \times \mathbb{R}_+ \to \mathbb{R}^3$ while $p : \Omega \times \mathbb{R}_+ \to \mathbb{R}_+$ is the pressure field. An hyperelastic two-dimensional (2D) closed structure is transported in such a fluid while deforming under hydrodynamics solicitations and responding to a constitutive equation. Let $\mathcal{M}(t) \subset \Omega$ be a 2D closed smooth immersed manifold for $t > 0$. The function $\mathbf{s} : \bigcup_{t\geq 0} \mathcal{M}(t) \times \{t\} \to \mathbb{R}^3$ returns the position of all points in the manifold current configuration. Such immersed manifold is endowed by the material density, elastic modulus and flexural rigidity, $\rho_s > 0$, $E > 0$, and $G > 0$, respectively. Therefore, the evolution of the material body within the surrounding fluid is ruled by the following initial-boundary value problem:

$$\begin{cases} \nabla \cdot \mathbf{u}(\mathbf{x}, t) = 0, & \mathbf{x} \in \Omega, t > 0 \\ \partial_t \mathbf{u}(\mathbf{x}, t) + (\mathbf{u}(\mathbf{x}, t) \cdot \nabla)\mathbf{u}(\mathbf{x}, t) = -\frac{1}{\rho_f}\nabla p(\mathbf{x}, t) + \nu \nabla^2 \mathbf{u}(\mathbf{x}, t), & \mathbf{x} \in \Omega, t > 0 \\ \mathbf{u}(\mathbf{x}, t) = \mathbf{u}_w(\mathbf{x}), & \mathbf{x} \in \partial\Omega, t \geq 0 \\ \mathbf{u}(\mathbf{x}, 0) = \mathbf{u}_0(\mathbf{x}), \; p(\mathbf{x}, 0) = p_0(\mathbf{x}), & \mathbf{x} \in \Omega, t = 0 \\ \rho_s \partial_{tt} \mathbf{s}(\mathbf{x}, t) - \mathbf{K}(\mathbf{s}(\cdot, t))(\mathbf{x}) = -\nabla p(\mathbf{x}, t) + \rho_f \nu \nabla^2 \mathbf{u}(\mathbf{x}, t), & \mathbf{x} \in \mathcal{M}(t), t > 0 \\ \mathbf{u}(\mathbf{x}, t) = \partial_t \mathbf{s}(\mathbf{x}, t), & \mathbf{x} \in \mathcal{M}(t), t > 0 \\ \mathbf{s}(\mathbf{x}, 0) = \mathbf{s}_0(\mathbf{x}), \; \partial_t \mathbf{s}(\mathbf{x}, 0) = \mathbf{u}_0(\mathbf{x}), & \mathbf{x} \in \mathcal{M}(0), t = 0 \end{cases}, \tag{13}$$

where $\mathbf{u}_w$, $\mathbf{u}_0$, and p_0 are functions ruling boundary and initial conditions for the fluid velocity and pressure. The material body responds to an elastic strain potential ($V^s(\mathbf{s}(\mathbf{x}, t))$) and a bending resistance ($V^b(\mathbf{s}(\mathbf{x}, t))$). As such, for every $t > 0$ the internal force density in $\mathbf{x} \in \mathcal{M}(t)$ is

$$\mathbf{K}(\mathbf{s}(\cdot, t))(\mathbf{x}) = \nabla \cdot (\nabla_{\mathbf{s}}(V^s(\mathbf{s}(\mathbf{x}, t)) + V^b(\mathbf{s}(\mathbf{x}, t)))\,. \tag{14}$$

with

$$V^s(\mathbf{s}(\mathbf{x}, t)) = \frac{1}{2} E \mathbf{s}^2(\mathbf{x}, t)\,,$$

$$V^b(\mathbf{s}(\mathbf{x}, t)) = \frac{1}{2} G \nabla^2(\mathbf{s}(\mathbf{x}, t))\,.$$

3.1 Fluid Evolution on Ω

In the lattice Boltzmann method, the solution of the continuum Boltzmann equation $\mathbf{f}(\mathbf{v}, \mathbf{x}, t)$ governing the probability of finding a particle (of fluid, in this context) with velocity $\mathbf{v}$ in $\mathbf{x} \in \Omega$ at the time $t > 0$ is projected onto a lattice endowed with a discrete set of 19 reticular velocities thus reducing $\mathbf{f}(\mathbf{v}, \mathbf{x}, t)$ to $\{\mathbf{f}_i(\mathbf{x}, t),\ i = 0, \cdots, 18\}$. As such Ω is approximated by a Cartesian grid O with uniform spacing Δx, Δt is the marching time step, and for $t > 0$ all nodes of the Cartesian grid $\mathbf{x} \in O$ are supplied with such discrete set of velocities. Let $\mathbf{e_i} \in \{-1, 0, 1\} \times \{-1, 0, 1\} \times \{-1, 0, 1\}, \forall i \in \{0, \cdots, 18\}$ be the 19 components of the $\mathbb{R}^3$-valued vectors defining this set of discrete velocities:

$$\mathbf{e} = \begin{pmatrix} 0 & 1 & -1 & 0 & 0 & 0 & 0 & 1 & 1 & 1 & 1 & -1 & -1 & -1 & -1 & 0 & 0 & 0 & 0 \\ 0 & 0 & 0 & 1 & -1 & 0 & 0 & 1 & -1 & 0 & 0 & 1 & -1 & 0 & 0 & 1 & 1 & -1 & -1 \\ 0 & 0 & 0 & 0 & 0 & 1 & -1 & 0 & 0 & 1 & -1 & 0 & 0 & 1 & -1 & 1 & -1 & 1 & -1 \end{pmatrix}. \tag{15}$$

For $t > 0$ and $\mathbf{x} \in O$ the discrete distribution functions $f_i : \Omega \times \mathbb{R}^+ \to \mathbb{R}$, $i = 0, \cdots, 18$ are the components of the solution of the reticular Boltzmann equation,

$$f_i(\mathbf{x}+\mathbf{e_i}\Delta t, t+\Delta t) - f_i(\mathbf{x}, t) = -\frac{\Delta t}{\tau}[f_i(\mathbf{x}, t) - f_i^{eq}(\mathbf{x}, t)] + \Delta t \mathcal{F}_i(\mathbf{x}, t)\,, \tag{16}$$

where:

- τ is the relaxation time. The kinematic viscosity of the flow is defined in this framework as $\nu = c_s^2(\tau - \frac{\Delta t}{2})$ being $c_s = 1/\sqrt{3}$ the reticular sound speed.
- $f_i^{eq} : \Omega \times \mathbb{R}^+ \to \mathbb{R}$, $i \in 0, \cdots, 18$ are the discrete local equilibrium functions corresponding for a Newtonian fluid to the discretized Maxwell–Boltzmann distribution:

$$f_i^{eq}(\mathbf{x}, t) = \omega_i \rho_f \left[1 + \frac{1}{c_s^2}(\mathbf{e}_i \cdot \mathbf{u}(\mathbf{x}, t)) + \frac{1}{2c_s^4}(\mathbf{e_i} \cdot \mathbf{u}(\mathbf{x}, t))^2 - \frac{1}{2c_s^2}\mathbf{u}^2(\mathbf{x}, t)\right], \tag{17}$$

  with $\omega_0 = 1/3$, $\omega_{1-6} = 1/18$, and $\omega_{7-18} = 1/36$.
- $\mathcal{F}_i : \Omega \times \mathbb{R}^+ \to \mathbb{R}$, $i = 0, \cdots, 18$ is a forcing term introduced for entrusting boundary conditions on $\mathcal{M}(t)$ (see Sect. 3.4) through the force $\mathbf{f}_{ib} : \Omega \times \mathbb{R}^+ \to \mathbb{R}^3$ accounting for the presence of the immersed structure:

$$\mathcal{F}_i(\mathbf{x}, t) = \left(1 - \frac{1}{2\tau}\right)\omega_i \left[\frac{\mathbf{e_i} - \mathbf{u}(\mathbf{x}, t)}{c_s^2} + \frac{\mathbf{e_i} \cdot \mathbf{u}(\mathbf{x}, t)}{c_s^4}\mathbf{e_i}\right] \cdot \mathbf{f}_{ib}(\mathbf{x}, t)\,. \tag{18}$$

In this context, the fluid dynamics macroscopic conserved quantities are defined by the statistical moments of $f_i(\mathbf{x}, t)$ with respect to $\mathbf{e}_i$:

- the fluid density: $\rho_f(\mathbf{x}, t) = \sum_{i=0}^{18} f_i(\mathbf{x}, t)$,
- the fluid momentum: $\rho_f(\mathbf{x}, t)\mathbf{u}(\mathbf{x}, t) = \sum_{i=0}^{18} f_i(\mathbf{x}, t)\mathbf{e}_i + \frac{\Delta t}{2}\mathbf{f}_{ib}(\mathbf{x}, t)$,
- the fluid pressure: $p(\mathbf{x}, t) = c_s^2 \sum_{i=0}^{18} f_i(\mathbf{x}, t)$.

3.2 Dirichlet Boundary Condition on $\partial\Omega$

The Zou and He [50] bounce-back conditions are adopted in this paper to enforce Dirichlet boundary conditions on $\partial\Omega$. Zou and He proposed a sophisticated bounce-back procedure in which the unknown components of f_i are computed by imposing $\mathbf{u}_w(\mathbf{x}) = (u_w(\mathbf{x}), v_w(\mathbf{x}), w_w(\mathbf{x}))$ at the boundary nodes $\mathbf{x} = \mathbf{x}_w$. For the north wall (used here as an example) such conditions read:

$$\begin{cases} \rho_f(x_w, y_w, z_w) = \frac{1}{w_w+1}\Big[f_0 + f_1 + f_2 + f_3 + f_4 + f_7 + f_8 + f_{13} \\ \qquad\qquad + f_{14} + 2\Big(f_5 + f_9 + f_{11} + f_{16} + f_{18}\Big)\Big], \\ N_{z,x} = \frac{1}{2}\Big(f_1 + f_7 + f_{13} - (f_2 + f_{14} + f_8)\Big) - \frac{1}{3}\rho_f(x_w, y_w, z_w)u_w , \\ N_{z,y} = \frac{1}{2}\Big(f_3 + f_7 + f_{14} - (f_4 + f_{13} + f_8)\Big) - \frac{1}{3}\rho_f(x_w, y_w, z_w)v_w , \\ f_6 = f_5 - \frac{1}{3}\rho_f(x_w, y_w, z_w)w_w , \\ f_{15} = f_{16} + \frac{1}{6}(-w_w + u_w) - N_{z,x} , \\ f_{10} = f_9 + \frac{1}{6}(-w_w - u_w) + N_{z,x} , \\ f_{17} = f_{18} + \frac{1}{6}(-w_w + v_w) - N_{z,y} , \\ f_{12} = f_{11} + \frac{1}{6}(-w_w - v_w) + N_{z,y} . \end{cases} \tag{19}$$

3.3 Evolution of $\mathcal{M}(t)$

The manifold $\mathcal{M}(t)$ is spatially discretized by a triangular mesh $T(t)$ composed by nv nodes connected by ne edges. On this mesh, for $t > 0$, the semi-discrete approximation of the momentum balance at $\mathbf{x}_i \in T(t)$, $\forall i \in \{1, \cdots, nv\}$, reads:

$$m_i^s \ddot{\mathbf{s}}_i(t) - \mathbf{K}(\mathbf{s}_i)(t) = -(p_i^+(t) - p_i^-(t))\mathbf{n}_i A_i + (\overline{\tau}_i^+(t) - \overline{\tau}_i^-(t))\mathbf{n}_i A_i , \tag{20}$$

being

- m_i^s the mass associated to the selected vertex;
- $\overline{\tau}_i(t)$ the viscous stress tensor;
- $\mathbf{n}_i$ the outward normal unity vector;
- A_i the portion of the total surface of $\mathcal{M}(t)$ associated to i taken as one third of the surface of the triangular elements sharing the vertex.

To avoid numerical instabilities and leakages through T(t), pressure and velocity derivatives are computed as the balance of their values taken at a distance of $1.2\,\Delta x$ along the inward (−) and outward (+) normals. A moving least squares reconstruction is adopted for approximating the value of $p_i(t)$ and $\overline{\tau}_i(t)$ [37, 48, 49]. Namely, $p_i^{\pm}(t)$ and $\overline{\tau}_i^{\pm}(t)$ are reconstructed from a stencil of 27 points centered on $\mathbf{x}_i \in T(t)$ (see Sect. 3.4).

Bodies internal force $\mathbf{K}(\mathbf{s}_i)(t)$ is computed as the resultant of the elastic strain response (Eq. (22)), the bending resistance (Eq. (24)), and the enclosed volume conservation force (Eq. (25)) [49]. The stretching elastic potential acting on the two vertices sharing the l-th edge is given by

$$V_l^s = \frac{1}{2} k_{s,l} (l_l - l_{l,0})^2 , \tag{21}$$

being $k_{s,l}$ [N/m] the elastic modulus for the edge l, l_l its length in the deformed configuration and $l_{l,0}$ its length at the resting. The elastic forces for vertices 1 and 2 connected by the edge l are:

$$\begin{cases} \mathbf{F}_1^{int,s} &= -k_{s,l}(l - l_0)\frac{\mathbf{r}_{1,2}}{l} \\ \mathbf{F}_2^{int,s} &= -k_{s,l}(l - l_0)\frac{\mathbf{r}_{2,1}}{l} \end{cases} , \tag{22}$$

where $\mathbf{r}_{i,j} = \mathbf{r}_i - \mathbf{r}_j$ and $\mathbf{r}_i$ is the ray vector connecting i to j. The elastic modulus of the immersed surface is obtained trough the non-dimensional group measuring the ratio between viscous drag forces versus surface tension forces, the capillary number Ca= $\frac{\rho \nu u_{ref}}{k_{s,0}}$. Then, it is spread to all edges considering the fraction of the total surface assigned to each edge, namely the sum of the area of the two elements sharing the selected edge A_i^l : $k_{s,l} = \frac{k_{s,0} T \sum_i A_i^l}{l^2}$ [51]. ρ_f and ν are the fluid density and kinematic viscosity; u_{ref} the reference velocity, $k_{s,0}$ the surface elastic modulus, and T the surface thickness.

Off-plane deformations are constrained by a bending potential depending on the difference between the local curvature in the resting and the deformed configurations, $c_{v,0}$ and c_v, respectively,

$$V_v^b = \frac{1}{2} k_b (c_v - c_{v,0})^2 , \tag{23}$$

being k_b the bending modulus. c_v is computed through the angle defined by the unit normal vectors of two adjacent elements, $\mathbf{n}_1$ and $\mathbf{n}_2$, $(\theta - \theta_0)$, with θ_0 the angle in the resting configuration. k_b is obtained as a function of $k_{s,l}$ through the non-dimensional group $Eb = \frac{k_b}{k_{s,l} r^2}$. Tan et al. [52] The resulting force on the involved four nodes reads:

$$\begin{cases} \mathbf{F}_1^{int,b} = \beta_b \big[b_{11}(\mathbf{n}_1 \times \mathbf{r}_{32}) + b_{12}(\mathbf{n}_2 \times \mathbf{r}_{32}) \big] \\ \mathbf{F}_2^{int,b} = \beta_b \big[b_{11}(\mathbf{n}_1 \times \mathbf{r}_{13}) + b_{12}(\mathbf{n}_1 \times \mathbf{r}_{34} + \mathbf{n}_2 \times \mathbf{r}_{13}) + b_{22}(\mathbf{n}_2 \times \mathbf{r}_{34}) \big] \\ \mathbf{F}_3^{int,b} = \beta_b \big[b_{11}(\mathbf{n}_1 \times \mathbf{r}_{21}) + b_{12}(\mathbf{n}_1 \times \mathbf{r}_{42} + \mathbf{n}_2 \times \mathbf{r}_{21}) + b_{22}(\mathbf{n}_2 \times \mathbf{r}_{42}) \big] \\ \mathbf{F}_4^{int,b} = \beta_b \big[b_{11}(\mathbf{n}_1 \times \mathbf{r}_{23}) + b_{22}(\mathbf{n}_2 \times \mathbf{r}_{23}) \big] \end{cases}, \tag{24}$$

with $b_{ij} = \frac{\mathbf{n}_i \cdot \mathbf{n}_j}{|\mathbf{n}_i||\mathbf{n}_j|}$ and $\beta_b = k_b \frac{sin(\theta)cos(\theta_0) - cos(\theta)sin(\theta_0)}{\sqrt{1 - cos^2(\theta)}}$.

Although the conservation of the total fluid volume enclosed by the immersed structures is granted from the fluid incompressibility. The fluid incompressibility do not guarantee that the volume of each capsule is conserved, while it only constrain the total fluid volume enclosed by all capsules. Due to this reason, the incompressibility (conservation of the enclosed volume) of a membrane is enforced through an effective penalty model:

$$\mathbf{F}_i^{int,v} = -k_v \left(1 - \frac{V}{V_0}\right) \mathbf{n}_i A_i \,, \tag{25}$$

k_v an incompressibility constant; V and V_0 the current and initial volume enclosed by the structure; A_i the area associated with the element i.

The semi-discrete problem (20) is integrated in time by adopting a P-(EC)^k β-Newmark scheme:

- Predictor Step:

$$\begin{cases} \dot{\mathbf{s}}^P_{(i,n+1)} = \dot{\mathbf{s}}_{(i,n)} + (1-\gamma)\Delta t \ddot{\mathbf{s}}_{(i,n)} \\ \mathbf{s}^P_{(i,n+1)} = \mathbf{s}_{(i,n)} + \Delta t \dot{\mathbf{s}}_{(i,n)} + (1-2\beta)\frac{\Delta t^2}{2}\ddot{\mathbf{s}}_{(i,n)} \end{cases}; \tag{26}$$

- Corrector Step:
  – Temporarily enforcing no-slip conditions on $T(t)$:

$$\mathbf{u}^P_{i,n+1} = \dot{\mathbf{s}}^P_{i,n+1} \,; \tag{27}$$

  – Temporarily updating the fluid solution $\forall e \in O : \mathbf{u}_{e,n} \rightarrow \mathbf{u}^P_{e,n+1}$;
  – Computing internal and external stresses $\mathbf{K}(\mathbf{s}^P_{i,n+1})$, $\nabla p^P_{i,n+1}$, and $\rho_f \nu \nabla^2 \mathbf{u}^P_{i,n+1}$;

- Updating $\ddot{\mathbf{s}}_{(i,n)} \rightarrow \ddot{\mathbf{s}}_{(i,n+1)}$:

$$\ddot{\mathbf{s}}_{(i,n+1)} = \frac{\mathbf{K}(\mathbf{s}^P_{i,n+1}) - \nabla p^P_{i,n+1} + \rho_f \nu \nabla^2 \mathbf{u}^P_{i,n+1}}{\rho_s}; \tag{28}$$

- Updating $\mathbf{s}_{(i,n)} \rightarrow \mathbf{s}_{(i,n+1)}$ and $\dot{\mathbf{s}}_{(i,n)} \rightarrow \dot{\mathbf{s}}_{(i,n+1)}$:

$$\begin{cases} \dot{\mathbf{s}}_{(i,n+1)} = \dot{\mathbf{s}}^P_{(i,n+1)} + \gamma \Delta t \ddot{\mathbf{s}}_{(i,n+1)}, \\ \mathbf{s}_{(i,n+1)} = \mathbf{s}^P_{(i,n+1)} + \beta \Delta t^2 \ddot{\mathbf{s}}_{(i,n+1)}; \end{cases} \tag{29}$$

Indeed, being the external stresses taken by the surrounding fluid, a fake temporary instance of the fluid solution at the time level n is used for enforcing temporary no-slip conditions on $T(t)$ based on the predicted quantities $\mathbf{s}^P_{(i,n+1)}$ and $\dot{\mathbf{s}}^P_{(i,n+1)}$ and computing the unknown $\ddot{\mathbf{s}}_{(i,n)}$.

3.4 Dirichlet Boundary Conditions on $\mathcal{M}(t)$

No-slip boundary conditions on $\mathbf{x} \in T(t)$ are enforced by transferring a convenient body force term $\mathbf{f}_{ib}(\mathbf{x}, t)$ to the Eulerian grid. $\mathbf{f}_{ib}(\mathbf{x}, t)$ is proportional to the difference between the velocity $\dot{s}(\mathbf{x}_l, t)$ of a point in T(t) and the velocity $\mathbf{u}(\mathbf{x}_\ell, t)$ the fluid would have if no boundary was immersed in. Although $\dot{s}(\mathbf{x}_\ell, t)$ is known from the solution of Eq. (20), $\mathbf{u}_f(\mathbf{x}_\ell, t)$ is reconstructed by means of a moving least squares approximation on a support for each $\mathbf{x}_\ell \in T(t)$:

$$\mathbf{u}(\mathbf{x}_\ell, t) = \sum_{k=1}^{27} \phi^\ell_k \mathbf{u}(\mathbf{x}^k_\ell, t), \tag{30}$$

where $\mathbf{u}(\mathbf{x}^k_\ell, t)$ is fluid velocity taken at the k-th point in the support domain of $\mathbf{x}_\ell$ and ϕ^ℓ_k is the ℓ, k-th coefficient of the least squares approximation. The chosen support of the point $\mathbf{x}_\ell$ is the ensemble of 27 Eulerian point in the cube by side $r_w = 1.3\,\Delta x$ centered on $\mathbf{x}_\ell$, namely $\mathbf{x}^k_\ell \in [x_\ell - \frac{1}{2}r_w, x_\ell + \frac{1}{2}r_w] \times [y_\ell - \frac{1}{2}r_w, y_\ell + \frac{1}{2}r_w] \times [z_\ell - \frac{1}{2}r_w, z_\ell + \frac{1}{2}r_w]$. Consequently, the needed force due to the presence of $\mathbf{x}_\ell$ is:

$$\mathbf{F}(\mathbf{x}_\ell, t) = \frac{\dot{\mathbf{s}}(\mathbf{x}_\ell, t) - \mathbf{u}(\mathbf{x}_\ell, t)}{\Delta t}, \tag{31}$$

and consequently

$$\mathbf{f}_{ib}(\mathbf{x}^k_\ell, t) = \sum_\ell c_\ell \phi^l_k \mathbf{F}(\mathbf{x}_\ell, t), \tag{32}$$

where, indeed, the sum on ℓ includes all points in $T(t)$ having $\mathbf{x}_\ell^k$ in their support. c_ℓ is a scaling factor for imposing the equivalence of the total force acting on the fluid between $T(t)$ and O [53, 54].

3.5 Strength and Limitations of the Dynamic-IB Approach

For enlightening strength and limitations of the proposed approach, we first review the analyses detailed in [37] and [49] for the measure of the deformation of a spherical membrane in a linear laminar flow in two- and three-dimensions, respectively. As a measure of the body deformation, the Taylor parameter is used:

$$D_{ij}(t) = \frac{M_{ij}(t) - m_{ij}(t)}{M_{ij}(t) + m_{ij}(t)},$$

where M and m are the largest and smallest diameters taken on the plane identified by the i-th and j-th directions. The dynamic-IB is critically compared to the more classic formulation—referenced as *kinematic-IB*—in which the deformation of the structure do not derive from the integration of the Newton law of motion, on the contrary, the velocity of the immersed manifold points is advected from the fluid. These two procedures were challenged by analyzing the obtained transient deformation of a capsule immersed in a shear flow for different mechanical stiffness and bending resistance values. The material body mechanical properties are characterized by two non-dimensional numbers, the dimensionless shear rate $G = \frac{\nu\rho_f\gamma 2r}{k_s}$ and the bending parameter $E_b = \frac{k_b}{k_s r^2}$, being γ the shear rate ($= \frac{2u_w}{H}$) and u_w the module of the velocity imposed at the top and bottom walls. The Reynolds number corresponds to Re = 0.05 and the density of the solid boundary is equal to the fluid one. For the 2D case, the accuracy of both formulations is measured with respect to the data published in [55]. As documented in Fig. 3, an excellent agreement is found with both models, especially for small G. However, the computational time needed by the dynamic-IB to perform 10,000 time steps was about 7% higher than that required by his kinematic counterpart. In [49] the three-dimensional system is studied (see Fig. 4). In this case, the agreement between the two formulations and the benchmark data [56] is confirmed as excellent. However, as previously observed by Walter et al. [57], a capsule with nearly zero thickness immersed in a linear flow will present corrugations near its equator region due to the compression generated as a response to shear-induced stretching. Specifically, the membrane section lying on the vorticity plane will stretch finding a compromise between hydrodynamic and internal stresses, while off-vorticity sections respond with compression. This mechanism brings to the well-known *membrane buckling* phenomenon. Interestingly, solely the dynamic-IB is able to

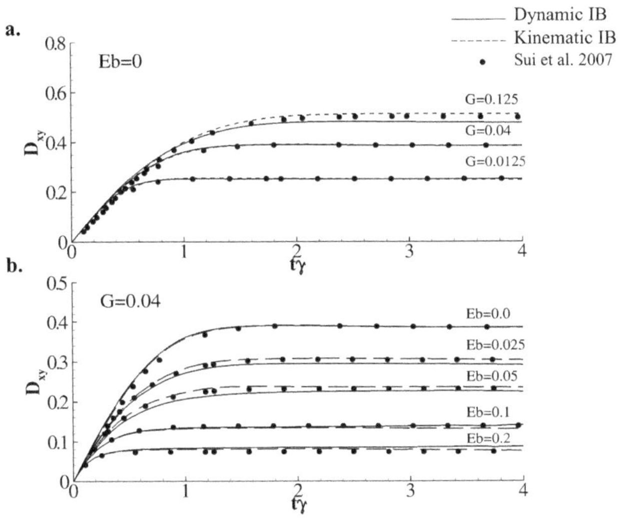

Fig. 3 Deformation of a circular capsule in shear flow (2D case) [37]. Variation of the Taylor parameter over time as function of the dimensionless shear rate G for Eb = 0 (**a**) and as function of the bending stiffness Eb for G = 0.04 (**b**)

reproduce the membrane buckling in agreement with the computations performed by Lac et al. [56] and the experimental findings by Walter et al. [57].

Note that, in the dynamic-IB approach the constitutive model for the immersed bodies is arbitrary. As the matter of fact, in [6] a model for decorating the immersed boundary surface with adhesive ligands was proposed and validated (also against experimental findings). Moreover, in [8] the interaction between circulating cells (by different shape) and adhesive microparticles (with different stiffness and adhesive coating) was studied (see Fig. 5). Specifically, it is observed that stiffer particles of any aspect ratio would be immediately dislodged along the capillary by the circulating cells. On the other side, softer particles would deform and convert a fraction of the momentum exchanged during the interaction with the circulating cells into deformation energy, thus reducing the forces directly exerted by the adhesive bonds. Due to this, soft and thin particles are preferred as micro-carriers in biomedical applications. Indeed, soft and thin carriers are able to firmly adhere to the vascular walls allowing for the sustained release of drugs from the lumen toward the diseased tissue. The dynamic-IB model is challenged forward by studying the collective behavior of particles (with different mechanical stiffness) transported within the red blood cell stream in physiological hematocrit [10]. Specifically, five independent initial configurations composed of rigid and soft particles within four different red blood cell distributions are considered thus identifying the collective behaviors averaged on the combination of those initial configurations. Firstly, it is measured the number of particles in time populating the three fluid regions generated by the cell stream, namely, the bulk zone, the

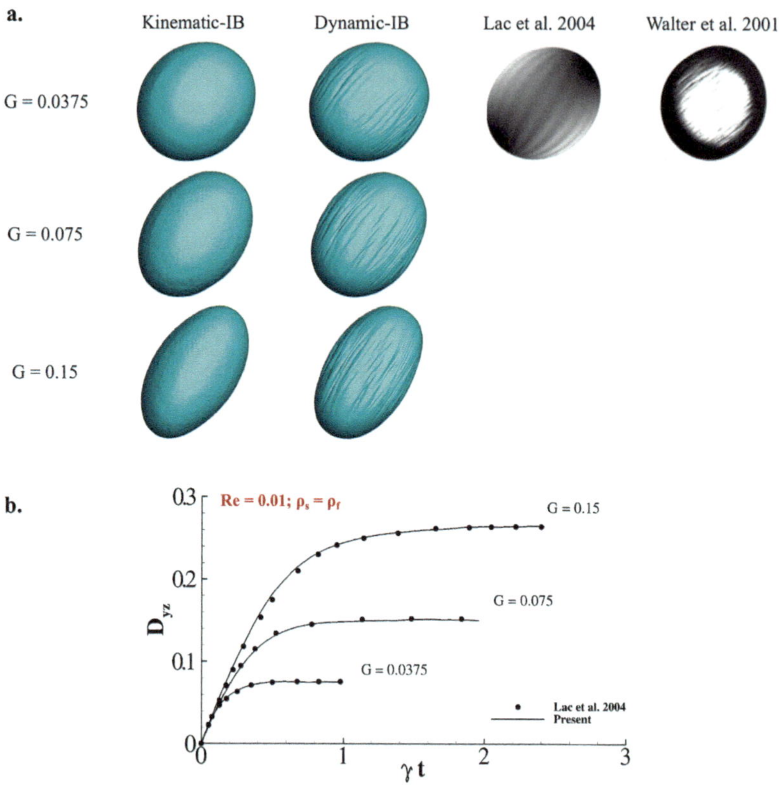

Fig. 4 Deformation of a circular capsule in shear flow (3D case) [49]. (**a**) Configurations of circular capsules for Eb = 0 and G = 0.0375, 0.075, and 0.15 obtained with kinematic-IB and dynamic-IB as compared with computations by Lac et al. [56] and experimental observation by Walter et al. [57]. (**b**) Distributions of the Taylor parameter over time as function of the dimensionless shear rate G (Eb = 0)

intermediate zone, and the cell-free layer. As demonstrated in Fig. 6a, rigid particles tend to marginate (reaching the cell-free layer) in the same fashion as their soft counterpart. Nonetheless, soft particles can adhere to the lumen more efficiently than rigid spheres. In fact, only 6% of the total number of rigid particles can firmly adhere to the vasculature while this number grows up to 23% when considering soft membranes (see Fig. 6b). In narrow capillaries, particle stiffness does not affect the margination rate. On the contrary, particle stiffness strongly affects the ability to efficiently interact with the endothelium and firmly adhere to walls.

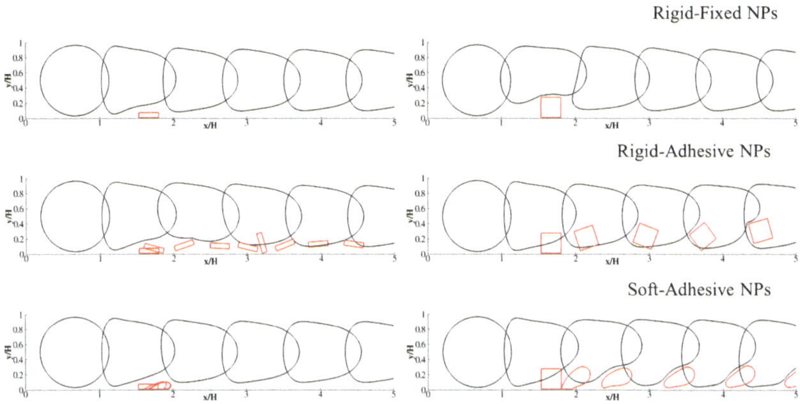

Fig. 5 Crawling of a cell over a micro-particle [8]. Crawling of a 7 μm cell crawling on a micro-particle by the height of 0.5 μm (left column) and 2.0 μm (right column) as a function of the mechanical stiffness

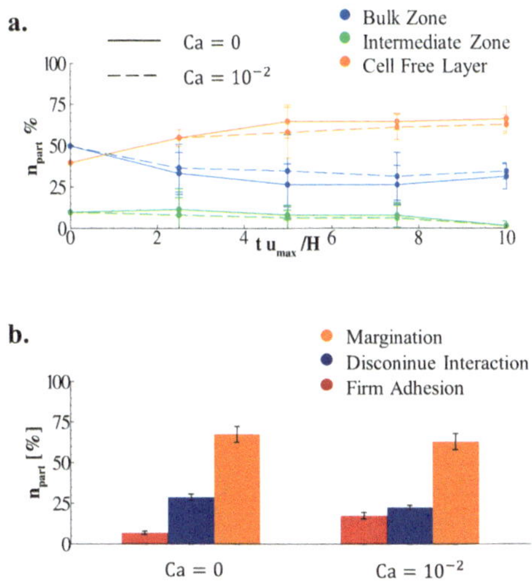

Fig. 6 Vascular fate of micro-particles in narrow capillaries [10]. (**a**) Distributions averaged over the red blood cell initial configurations of the number of particles over time populating the three flow regions. (**b**) Average number of particles firmly adhering the lumen, discontinuously interacting with the vasculature and marginating

4 Conclusive Remarks

In this contribution two emerging strategies for computing multiphase flows in the microcirculation regimen are presented and analyzed. Both models demonstrate promising features in term of numerical accuracy, moderate computational cost and applicability to modern physics problems. Moreover, it has been thoroughly presented the rigorous mathematical background supporting the two complex computational frameworks. Immersed boundary methods are computationally less expansive since they do not introduce additional Eulerian fields and are the preferred choice for simulations which do not involve topological changes of the interface; on the contrary, in phenomena where merging or break up could take place, the Volume of Fluid approach is preferred, since it inherently handle interface changes. Future works should focus on developing algorithms to combine the two methods such that the fluid-fluid interface can interact with the solid-fluid interface (contact line models).

Acknowledgments AC is a member of the Gruppo Nazionale per il Calcolo Scientifico (GNCS) of Istituto Nazionale di Alta Matematica (INdAM) and acknowledges the "INdAM—GNCS Project" (CUP—E53C22001930001). This work was partially supported by the project "Research for Innovation" (REFIN)—POR Puglia FESR FSE 2014-2020—Asse X—Azione 10.4 (CUP—D94120001410008).

References

1. Whitesides, G.M.: The origins and the future of microfluidics. Nature **442**(7101), 368–373 (2006)
2. Ganán-Calvo, A.M., Gordillo, J.M.: Perfectly monodisperse microbubbling by capillary flow focusing. Phys. Rev. Lett. **87**(27), 274501 (2001)
3. Tan, Y.-C., Fisher, J.S., Lee, A.I., Cristini, V., Lee, A.P.: Design of microfluidic channel geometries for the control of droplet volume, chemical concentration, and sorting. Lab Chip **4**(4), 292–298 (2004)
4. Bringer, M.R., Gerdts, C.J., Song, H., Tice, J.D., Ismagilov, R.F.: Microfluidic systems for chemical kinetics that rely on chaotic mixing in droplets. Philos. Trans. R. Soc. Lond. A Math. Phys. Eng. Sci. **362**(1818), 1087–1104 (2004)
5. Volpatti, L.R., Yetisen, A.K.: Commercialization of microfluidic devices. Trends Biotechnol. **32**(7), 347–350 (2014)
6. Coclite, A., Mollica, H., Ranaldo, S., Pascazio, G., de Tullio, M.D., Decuzzi, P.: Predicting different adhesive regimens of circulating particles at blood capillary walls. Microfluidics Nanofluidics **21**(11), 168 (2017). ISSN 1613-4990. https://doi.org/10.1007/s10404-017-2003-7
7. Mollica, H., Coclite, A., Miali, M.E., Pereira, R.C., Paleari, L., Manneschi, C., DeCensi, A., Decuzzi, P.: Deciphering the relative contribution of vascular inflammation and blood rheology in metastatic spreading. Biomicrofluidics **12**(4), 042205 (2018). https://doi.org/10.1063/1.5022879
8. Coclite, A., Pascazio, G., de Tullio, M.D., Decuzzi, P.: Predicting the vascular adhesion of deformable drug carriers in narrow capillaries traversed by blood cells. J. Fluids Struct. **82**, 638–650 (2018). ISSN 0889-9746. https://doi.org/10.1016/j.jfluidstructs.2018.08.001
9. Laurino, F., Coclite, A., Tiozzo, A., Decuzzi, P., Zunino, P.: A hierarchical multiscale model for predicting the vascular behavior of blood-borne nanomedicines. Int. J. Multiscale Comput. Eng. **18**(3), 335–359 (2020). https://doi.org/10.1615/IntJMultCompEng.2020033358
10. Coclite, A.: Vascular journey and adhesion mechanics of micro-sized carriers in narrow capillaries. Microvascular Res. **132**, 104069 (2020). ISSN 0026-2862. https://doi.org/10.1016/j.mvr.2020.104069
11. Perdigones, F.: Lab-on-PCB and flow driving: a critical review. Micromachines **12**(2), 175 (2021)

12. Gossett, D.R., Weaver, W.M., Mach, A.J., Hur, S.C., Tse, H.T.K., Lee, W., Amini, H., Di Carlo, D.: Label-free cell separation and sorting in microfluidic systems. Anal. Bioanal. Chem. **397**, 3249–3267 (2010)
13. Schramm, L.L.: Emulsions, Foams, Suspensions, and Aerosols: Microscience and Applications. Wiley, Hoboken (2014)
14. Tanner, R.I.: Aspects of non-colloidal suspension rheology. Phys. Fluids **30**(10), 101301 (2018)
15. Petsev, D.N.: Emulsions: Structure, Stability and Interactions. Elsevier, Amsterdam (2004)
16. Einstein, A.: Eine neue bestimmung der moleküldimensionen. Ph.D. Thesis, ETH Zurich (1905)
17. Einstein, A.: Berichtigung zu meiner arbeit: Eine neue bestimmung der moleküldimensionen. Ann. Phys. **339**(3), 591–592 (1911)
18. Batchelor, G.K., Green, J.T.: The hydrodynamic interaction of two small freely-moving spheres in a linear flow field. J. Fluid Mech. **56**(2), 375–400 (1972)
19. von Eilers, H.: Die viskosität von emulsionen hochviskoser stoffe als funktion der konzentration. Kolloid-Zeitschrift **97**(3), 313–321 (1941)
20. Mari, R., Seto, R., Morris, J.F., Denn, M.M.: Shear thickening, frictionless and frictional rheologies in non-brownian suspensions. J. Rheol. **58**(6), 1693–1724 (2014)
21. Guazzelli, É., Pouliquen, O.: Rheology of dense granular suspensions. J. Fluid Mech. **852**, P1 (2018)
22. Pries, A.R., Secomb, T.W., Gaehtgens, P.: Biophysical aspects of blood flow in the microvasculature. Cardiovascular Res. **32**(4), 654–667 (1996)
23. Taylor, G.I.: The viscosity of a fluid containing small drops of another fluid. Proc. R. Soc. Lond. A Contain. Pap. Math. Phys. Character **138**(834), 41–48 (1932)
24. Pal, R.: Viscous behavior of concentrated emulsions of two immiscible newtonian fluids with interfacial tension. J. Colloid Interface Sci. **263**(1), 296–305 (2003)
25. Oliveira, T.F., Cunha, F.R.: Emulsion rheology for steady and oscillatory shear flows at moderate and high viscosity ratio. Rheol. Acta **54**(11–12), 951–971 (2015)
26. Matsunaga, D., Imai, Y., Yamaguchi, T., Ishikawa, T.: Rheology of a dense suspension of spherical capsules under simple shear flow. J. Fluid Mech. **786**, 110–127 (2016)
27. Rosti, M.E., Brandt, L., Mitra, D. (2018). Rheology of suspensions of viscoelastic spheres: deformability as an effective volume fraction. Phys. Rev. Fluids **3**(1), 012301 (2018)
28. D'Avino, G., Maffettone, P.L.: Particle dynamics in viscoelastic liquids. J. Non-Newtonian Fluid Mech. **215**, 80–104 (2015)
29. Villone, M.M., d'Avino, G., Hulsen, M.A., Maffettone, P.L.: Dynamics of prolate spheroidal elastic particles in confined shear flow. Phys. Rev. E **92**(6), 062303 (2015)
30. Abtahi, S.A., Elfring, G.J.: Jeffery orbits in shear-thinning fluids. Phys. Fluids **31**(10), 103106 (2019)
31. Chesters, A.K.: Modelling of coalescence processes in fluid-liquid dispersions: a review of current understanding. Chem. Eng. Res. Des. **69**(A4), 259–270 (1991)
32. Guido, S., Simeone, M.: Binary collision of drops in simple shear flow by computer-assisted video optical microscopy. J. Fluid Mech. **357**, 1–20 (1998)
33. Sibillo, V., Pasquariello, G., Simeone, M., Cristini, V., Guido, S.: Drop deformation in microconfined shear flow. Phys. Rev. Lett. **97**(5), 054502 (2006)
34. Caserta, S., Guido, S.: Vorticity banding in biphasic polymer blends. Langmuir **28**(47), 16254–16262 (2012)
35. De Vita, F., Rosti, M.E., Caserta, S., Brandt, L.: On the effect of coalescence on the rheology of emulsions. J. Fluid Mech. **880**, 969–991 (2019)
36. Rosti, M.E., De Vita, F., Brandt, L.: Numerical simulations of emulsions in shear flows. Acta Mech. **230**(2), 667–682 (2019)
37. Coclite, A., Ranaldo, S., de Tullio, M.D., Decuzzi, P., Pascazio, G.: Kinematic and dynamic forcing strategies for predicting the transport of inertial capsules via a combined lattice Boltzmann immersed boundary method. Comput. Fluids **180**, 41–53 (2019). ISSN 0045-7930. https://doi.org/10.1016/j.compfluid.2018.12.014

38. Scardovelli, R., Zaleski, S.: Direct numerical simulation of free-surface and interfacial flow. Annu. Rev. Fluid Mech. **31**(1), 567–603 (1999)
39. Popinet, S.: An accurate adaptive solver for surface-tension-driven interfacial flows. J. Comput. Phys. **228**(16), 5838–5866 (2009)
40. Xiao, F., Honma, Y., Kono, T.: A simple algebraic interface capturing scheme using hyperbolic tangent function. Int. J. Numer. Methods Fluids **48**(9), 1023–1040 (2005)
41. Ii, S., Sugiyama, K., Takeuchi, S., Takagi, S., Matsumoto, Y., Xiao, F.: An interface capturing method with a continuous function: the thinc method with multi-dimensional reconstruction. J. Comput. Phys. **231**(5), 2328–2358 (2012)
42. Xiao, F., Ii, S., Chen, C.: Revisit to the thinc scheme: a simple algebraic vof algorithm. J. Comput. Phys. **230**(19), 7086–7092 (2011)
43. Kim, J., Moin, P.: Application of a fractional-step method to incompressible navier-stokes equations. J. Comput. Phys. **59**(2), 308–323 (1985)
44. Dodd, M.S., Ferrante, A.: A fast pressure-correction method for incompressible two-fluid flows. J. Comput. Phys. **273**, 416–434 (2014)
45. De Vita, F., De Lillo, F., Verzicco, R., Onorato, M.: A fully eulerian solver for the simulation of multiphase flows with solid bodies: application to surface gravity waves. J. Comput. Phys. **438**, 110355 (2021)
46. Sheth, K.S., Pozrikidis, C.: Effects of inertia on the deformation of liquid drops in simple shear flow. Comput. Fluids **24**(2), 101–119 (1995)
47. De Vita, F., Rosti, M.E., Caserta, S., Brandt, L.: Numerical simulations of vorticity banding of emulsions in shear flows. Soft Matter **16**(11), 2854–2863 (2020)
48. Coclite, A., de Tullio, M.D., Pascazio, G., Decuzzi, P.: A combined lattice Boltzmann and immersed boundary approach for predicting the vascular transport of differently shaped particles. Comput. Fluids **136**, 260–271 (2016). ISSN 0045-7930. http://dx.doi.org/10.1016/j.compfluid.2016.06.014
49. Coclite, A., Ranaldo, S., Pascazio, G., de Tullio, M.D.: A lattice boltzmann dynamic-immersed boundary scheme for the transport of deformable inertial capsules in low-re flows. Comput. Math. Appl. **80**(12), 2860–2876 (2020). ISSN 0898-1221. https://doi.org/10.1016/j.camwa.2020.09.017
50. Zou, Q., He, X.: On pressure and velocity boundary conditions for the lattice Boltzmann BGK model. Phys. Fluids **9**(6), 1591–1598 (1997). http://dx.doi.org/10.1063/1.869307
51. Van Gelder, A.: Approximate simulation of elastic membranes by triangulated spring meshes. J. Graph. Tools **3**(2), 21–41 (1998)
52. Tan, J., Thomas, A., Liu, Y.: Influence of red blood cells on nanoparticle targeted delivery in microcirculation. Soft Matter **8**(6), 1934–1946 (2012). https://doi.org/10.1039/c2sm06391c
53. Vanella, M., Balaras, E.: A moving-least-squares reconstruction for embedded-boundary formulations. J. Comput. Phys. **228**(18), 6617–6628 (2009). ISSN 0021-9991. http://dx.doi.org/10.1016/j.jcp.2009.06.003
54. de Tullio, M.D., Pascazio, G.: A moving-least-squares immersed boundary method for simulating the fluid-structure interaction of elastic bodies with arbitrary thickness. J. Comput. Phys. **35**, 201–225 (2016)
55. Sui, Y., Chew, Y.T., Roy, P., Chen, X.B., Low, H.T.: Transient deformation of elastic capsules in shear flow: effect of membrane bending stiffness. Phys. Rev. E **75**(6), 066301 (2007)
56. Lac, E., Barthes-Biesel, D., Pelekasis, N.A., Tsamopoulos, J.: Spherical capsules in three-dimensional unbounded Stokes flows: effect of the membrane constitutive law and onset of buckling. J. Fluid Mech. **516**, 303–334 (2004)
57. Walter, A., Rehage, H., Leonhard, H.: Shear induced deformation of microcapsules: shape oscillations and membrane folding. Colloids Surfaces A Physicochem. Eng. Aspects **183**, 123–132 (2001)

Application of the Zenger Correction to an Elliptic PDE with Dirac Source Term

Tobias Köppl

Abstract In this work, a numerical solution technique for an elliptic Dirac problem is investigated. One reason to study this issue is that this partial differential equation (PDE) is closely related to a mathematical model for flow processes in vascularized tissue. In order to circumvent the numerical difficulties associated with a Dirac source term a regularization technique is used, which is known as Zenger correction. Using the Zenger correction, it can be shown that a Finite Difference discretization for a standard Dirac problem converges with optimal order in regions excluding the singularity of the solution. Besides a Dirac source term, a coupling term occurring in the right hand side of the equation has to be handled. For a numerical treatment of this coupling term an iterative solution scheme is proposed. Thereby, in each iteration step a standard elliptic Dirac problem is solved. The performance of the newly developed numerical solver is illustrated by means of a series of numerical tests.

Keywords Dirac problems · 3D-1D coupled problems · Zenger correction · Regularizations · Local convergence

1 Introduction

Mathematical modeling of flow and transport processes in vascularized tissue has been objective of several publications in the past decades [1, 2]. This is due to the fact that by means of a mathematical model diseases and therapies can be studied in a non-invasive way. One example is the simulation of tumor growth [3, 4] and therapies such as hyperthermia [5].

It is obvious that modeling flow in vascularized tissue is challenging due to the fact that flow in a blood vessel network consisting of a huge number of vessels has to

T. Köppl (✉)
Hasselt University, Diepenbeek, Belgium
e-mail: tobias.koeppl@uhasselt.be

A. Linninger et al. (eds.), *Quantitative Approaches to Microcirculation*,
SEMA SIMAI Springer Series 36, https://doi.org/10.1007/978-3-031-58519-7_8

be modeled [6]. Quite often one is confronted with the fact that the vessel geometry can only be reconstructed from images with huge segmentations errors due to noise. To tackle this problem the missing geometry can be generated in an artificial way [7–10]. Another difficulty is that the interaction of network flow and interstitial flow has to be taken into account. For this purpose, the porous structure of tissue supplied by the network has to resolved producing an enormous amount of data.

One way to decrease the high amount of data and the numerical effort is to consider tissue as a three-dimensional (3D) homogenized porous medium in which flow is governed by Darcy's equation [3, 11, 12]. The blood vessel network embedded in the respective tissue block is reduced to a one-dimensional (1D) graph-like structure [13]. By this a complicated meshing of the network is avoided. Within a single vessel, flow is governed by a 1D Hagen-Poiseuille type flow. To ensure mass conservation in the whole network, the different flow problems connected at a branching point are coupled such that conservation of mass is guaranteed [14]. Now the issue arises, how the flow processes in the 1D network and the 3D porous tissue can be coupled. A possible approach is based on Starling's filtration law describing flow through a permeable membrane. Starling's filtration law determines the source terms in both PDEs, where in Darcy's equation a Dirac measure occurs that is concentrated on the centerlines of the network. All in all, this results in a 3D-1D coupled elliptic PDE system [15].

However, the simplification processes resulting in a 3D-1D coupled PDE system has the drawback that along the centerline of the network singularities in the 3D solution are occurring. This causes difficulties both in terms of an analytical and numerical treatment. It can be shown that the solution of the 3D problem is not contained in an H^1-space [16]. To prove the well-posedness of the 3D-1D coupled problem, weighted Sobolev spaces have to be introduced [17].

The reduced regularity of the 3D solution deteriorates the convergence behavior of numerical solution schemes like standard Finite Element methods (FEM) [18]. In the recent decades several numerical solution strategies have been developed. Optimal convergence of standard FEM can be shown, if weighted Sobolev spaces and graded meshes are used to resolve the singularity [17, 19]. Graded meshes are triangular meshes refined around the singularity. They are obtained from uniform triangular meshes by scaling the diameter of the triangles with respect to the distance to the singularity. If either graded meshes or weighted Sobolev spaces are not considered, quasi-optimal convergence of FEM can be shown [20, 21]. However, graded meshes make the handling of network structures very complicated, since the generation of the 3D mesh discretizing the tissue domain is not trivial. Due to that further approaches have been developed such as the singularity removal method proposed in [22]. Thereby, the singular components of the solution are represented by a combination of Green's functions, while the smooth components are determined by means of standard elliptic problems with L^2-right hand sides. The advantage of this approach is that structured and relatively coarse meshes can be used contrary to graded meshes. Other investigations are not focused on global discretization errors but on local discretization errors neglecting a region around the singularities. Since the solution is usually smooth at a distance from the singularities optimal convergence of standard FEM and discontinuous Galerkin methods (DG) can be shown [23, 24]. Another approach yielding locally a precise approximation of the solution as well as the volume flux between the vascular system and the interstitial space is based on replacing the Dirac measure by a smooth Kernel function [25–27].

In this work, the approach based on approximating the Dirac measure by means of a smooth Kernel function is reconsidered. Contrary to other Kernel methods, we use a Kernel that is referred to as Zenger correction [28, 29]. The idea is to integrate the Heaviside function several times until a sufficiently smooth function is obtained. Deriving this function numerically by means of Finite Differences an appropriate Kernel function results. It can be shown that a Finite Difference discretization applied to a standard Dirac problem converges locally with optimal order. If the Kernel is sufficiently smooth, higher order approximations can be obtained by means of extrapolation [30]. An advantage of this method is that a higher order approximation of the solution with compact discretization stencils is constructed. However, up to now this technique has only been applied to Dirac problems without coupling terms that are occurring in terms of the 3D-1D coupled flow problem. Thus a fixed point iteration scheme as in [31] is applied such that in each iteration a decoupled Dirac problem can be solved. Thereby, it is not necessary to include the coupling term into the system matrix of the discrete system.

The remainder of this work is structured as follows: In Sect. 2, we introduce a two-dimensional (2D) model problem that is used to illustrate the basic ideas of our numerical solution method. The following two sections are concerned with the fundamentals of the Zenger correction and the numerical treatment of the model problem from Sect. 2. In order to illustrate the performance of the new algorithm, a series of numerical tests is performed in Sect. 5. Finally, in Sect. 6 some concluding remarks and an outlook are provided.

2 Model Problem

In context with dimensional reduced modeling of flow processes in vascularized tissue the following PDE system has been investigated in a series of publications [13, 15]:

$$\begin{aligned} -\nabla \cdot (K_{3D} \nabla u_{3D}) &= 2\pi R L_p \left(u_{1D} - \overline{u}_{3D} \right) \delta_\Lambda, \text{ in } \Omega, \\ -\frac{\partial}{\partial s}\left(K_{1D} \frac{\partial u_{1D}}{\partial s} \right) &= 2\pi R L_p \left(\overline{u}_{3D} - u_{1D} \right), \text{ in } \Lambda. \end{aligned} \tag{1}$$

$\Omega \subset \mathbb{R}^3$ is a compact polyhedral domain representing the tissue matrix with a permeability tensor K_{3D}. $\Lambda \subset \Omega$ is the centerline of a vessel with radius R

supplying the tissue domain Ω. It is parameterized by a curve parameter s:

$$\Lambda : [0,1] \rightarrow \Omega, \; s \mapsto \Lambda(s).$$

Thereby, we assume that Λ is a sufficiently smooth curve and not intersecting. The permeability of the vessel wall is given by L_p and K_{1D} is the permeability of the vessel lumen. Finally, the solution variables u_{3D} and u_{1D} represent the pressures in the tissue and the blood vessel. In order to be able to compute a difference between the tissue and vessel pressure, the tissue pressure has to be averaged and reduced to a 1D quantity that is denoted by $\overline{u}_{3D}$. For a curve parameter $s \in [0,1]$ it is computed as follows:

$$\overline{u}_{3D}(s) = \frac{1}{2\pi} \int_0^{2\pi} u_{3D}\left(\Lambda(s) + R\mathbf{b}(s,\theta)\right) \, d\theta.$$

$\mathbf{b}(s,\theta), \; s \in [0,1]$ is a unit vector perpendicular to Λ in $\Lambda(s)$ rotating around Λ by an angle $\theta \in [0, 2\pi)$. A well-posed problem is obtained, by prescribing appropriate boundary conditions at the respective boundaries:

$$u_{1D}(0) = u_0, \; u_{1D}(1) = u_1 \text{ and } u_{3D}(x) = u_B(x), \; x \in \partial\Omega.$$

Instead of (1), we consider a simplified model problem in order to obtain first insights in the performance of the numerical solution method that is discussed in the upcoming sections. Designing a simpler model it is assumed that the 1D solution u_{1D} is fixed to a constant $U \in \mathbb{R}$: $u_{1D} \equiv U$. Moreover, the following the choices are made:

$$K_{3D} = I, \; u = u_{3D} \text{ and } K_{1D} = 1.$$

Assuming further that Λ is a straight line, it is sufficient to consider a 2D elliptic problem:

$$\begin{aligned} -\Delta u &= 2\pi R L_p \left(U - \overline{u}\right) \delta_{x_0}, \text{ in } \Omega, \\ u &= u_B, \text{ on } \partial\Omega. \end{aligned} \tag{2}$$

$\Omega \subset \mathbb{R}^2$ is a polyhedral domain and $x_0 \in \Omega$. The averaged quantity $\overline{u}$ is given by:

$$\overline{u} = \frac{1}{2\pi} \int_0^{2\pi} u\left(x_0 + R \begin{pmatrix} \cos(\theta) \\ \sin(\theta) \end{pmatrix}\right) \, d\theta.$$

This problem has already been considered in [17]. In the following, it is denoted as 2D-0D coupled problem, since we consider a two-dimensional problem and the vascular solution is given by a constant U.

The main motivation for first developing solution techniques in terms of (2) instead of (1) is that the latter problem is a fully coupled model i.e. the 1D solution defined on Λ influences the 3D solution on Ω and vice versa. Therefore, the numerical treatment of a Dirac measure in the source term and the average term $\bar{u}$ can not be investigated separately. Obviously (2) is decoupled, since the 1D solution u_{1D} is fixed to U. Thus, it is possible to study numerical solution techniques for both issues without the need to discretize the 1D problem. Moreover, in (2) one is confronted with a point source term, while in (1) a line source term has to be handled. According to [27], it is meaningful to first understand the numerical treatment of a point source and to extend this insights in a second step to a line source term by representing the line source term as a weighted sum of point sources. Finally, incorporating the term $\overline{u}$ into a numerical scheme is not trivial. Discretizing $\overline{u}$ e.g. by a standard FEM, one can easily see that the resulting system matrix deviates from the standard pattern of a FEM matrix and that on sufficiently fine meshes a matrix free implementation of the solver is hard to realize. A LU decomposition of the system matrix might cause significantly more fill ins compared to the LU decomposition of the standard FEM matrix. In terms of iterative solvers it is still discussed how multigrid methods perform and how a possible preconditioner has to be designed. An approach on how to determine efficient preconditioners for a 3D-1D coupled problem is discussed in [32]. The issue of deriving a full discretization of (1) and presenting application relevant simulation results is a matter of upcoming works.

3 Iterative Solution Scheme

As a first step towards a numerical solver, a strategy is developed how to handle the averaged value in the source term of (2). Thereby, it is intended to keep the discretization of the Laplace operator and the average, separately. One way to achieve this, is to consider a sequence of Dirac problems, given by:

$$\begin{aligned} -\Delta u^{(k+1)} &= 2\pi R L_p \left(U - \overline{u^{(k)}} \right) \delta_{x_0}, \text{ in } \Omega, \\ u^{(k+1)} &= u_B, \text{ on } \partial\Omega, \end{aligned} \tag{3}$$

where $k \in \mathbb{N} \cup \{0\}$. This means that in the $k+1$-st iteration a standard Dirac problem has to be solved. After solving this problem, the average in the source term is computed by means of an appropriate quadrature rule with weights ω_j and nodes $\theta_j \in [0, 1]$, $j = 1, \ldots, nQP$:

$$\overline{u^{(k)}} = \int_0^1 u^{(k)} \left(\begin{pmatrix} x_{0,1} \\ x_{0,2} \end{pmatrix} + R \cdot \begin{pmatrix} \cos(2\pi s) \\ \sin(2\pi s) \end{pmatrix} \right) ds \approx \sum_{j=1}^{nQP} \omega_j \cdot u^{(k)}(\xi_j). \tag{4}$$

Table 1 Discretization stencils for the Laplace operator using a mesh size h [33]

Name	Symbol	Order	Stencil
Five point	$-\Delta_h^{(5)}$	$\mathcal{O}(h^2)$	$\frac{1}{h^2}\begin{pmatrix} 0 & -1 & 0 \\ -1 & 4 & -1 \\ 0 & -1 & 0 \end{pmatrix}$
Nine point	$-\Delta_h^{(9)}$	$\mathcal{O}(h^4)$	$\frac{1}{6h^2}\begin{pmatrix} -1 & -4 & -1 \\ -4 & 20 & -4 \\ -1 & -4 & -1 \end{pmatrix}$

The nodes ξ_j are given by:

$$\xi_j = \underbrace{\begin{pmatrix} x_{0,1} \\ x_{0,2} \end{pmatrix}}_{=x_0} + R \cdot \begin{pmatrix} \cos(2\pi\theta_j) \\ \sin(2\pi\theta_j) \end{pmatrix}. \tag{5}$$

In order to solve (3) numerically, we use a Finite Difference method based on compact discretization stencils. The first one is a second order Five point and a fourth order Nine point discretization stencil for the Laplace operator (see Table 1). The corresponding matrices are denoted by $A_h^{(5)} \in \mathbb{R}^{N \times N}$ and $A_h^{(9)} \in \mathbb{R}^{N \times N}$, where $N \in \mathbb{N}$ denotes the number of grid nodes and h is mesh size of the computational grid. A grid node x_i contained in this grid is given by:

$$x_i = ((m-1) \cdot h, (n-1) \cdot h) = x_{m,n}, \; m, n \in \{1, \ldots, N_d\},$$
$$h = \frac{1}{N_d - 1}, \; N = N_d^2.$$

N_d represents the number of grid nodes in x or y direction, i.e., we restrict ourselves to uniform meshes consisting of squares and choose the unit square as a computational domain: $\Omega = (0, 1)^2$. The index i is given by:

$$i = n + (m-1) \cdot N_d.$$

$U_h^{(k+1)} \in \mathbb{R}^N$ is the vector of solution values. The i-th component contains the numerical approximation at the i-th grid node with respect to the solution $u^{(k+1)}$.

It remains to derive a numerical approximation of the Dirac source term δ_{x_0}. As it has already been mentioned, the Zenger correction is used to emulate the influence of δ_{x_0}. To increase the readability of the text, the discussion of the most important features of this regularization method is postponed to the upcoming section. Let us further assume that the vector $b_h \in \mathbb{R}^N$ contains the boundary conditions and the evaluations of the regularized Delta distribution. All in all we have to solve the following linear system equation of the $k+1$-st iteration:

$$A_h^{(l)} U_h^{(k+1)} = b_h, \tag{6}$$

where $l = 5$ or $l = 9$ holds. The vector $b_h \in \mathbb{R}^N$ is given by:

$$b_h(i) = \begin{cases} u_B(x_i), & \text{if } x_i \in \partial(0,1)^2, \\ \alpha_k \tilde{\delta}_{x_0}, & \text{if } x_i \in \operatorname{supp}\left(\tilde{\delta}_{x_0}\right), \end{cases} \tag{7}$$

where α_k is given by:

$$\alpha_k = 2\pi R L_p \left(U - \overline{U_h^{(k)}} \right). \tag{8}$$

For all the other nodes i, it holds that $b_h(i) = 0$. $\tilde{\delta}_{x_0}$ denotes a regularization of $\tilde{\delta}_{x_0}$ and $\operatorname{supp}\left(\tilde{\delta}_{x_0}\right)$ is its support. Thereby, it is assumed that h is sufficiently small such that

$$\operatorname{supp}\left(\tilde{\delta}_{x_0}\right) \cap \partial(0,1)^2 = \emptyset$$

is ensured. To enforce the boundary conditions, it has to hold for the entry (i, j), $x_i \in \partial(0,1)^2$ of the matrix $A_h^{(l)}$:

$$A_h^{(l)}(i, j) = \begin{cases} 1 \text{ if } i = j, \\ 0 \text{ if } i \neq j. \end{cases} \tag{9}$$

In case of the Nine point stencil, the right hand side has to be modified such that a convergence of fourth order can be observed. Considering the i-th component of b_h corresponding to an inner grid node

$$x_i = x_{m,n} = ((m-1) \cdot h, (n-1) \cdot h) \notin \partial(0,1)^2$$

one has to exert the following computation:

$$\begin{aligned} b_h(i) = b_h\left(x_{m,n}\right) + \frac{1}{12}\Big(&-b_h\left(x_{m,n-1}\right) + 4b_h\left(x_{m,n}\right) \\ &-b_h\left(x_{m,n+1}\right) - b_h\left(x_{m+1,n}\right) - b_h\left(x_{m-1,n}\right)\Big). \end{aligned} \tag{10}$$

For a boundary node $x_i \in \partial(0,1)^2$ it has to hold:

$$b_h(i) = b_h(x_i) = u_B(x_i). \tag{11}$$

Please note that the linear systems of equations can be solved by a classical iterative solver or a multigrid solver [33]. In this context the matrix $A_h^{(l)}$ does not have to be assembled, since for these solvers only matrix vector products have to be computed. Due to the fact that compact discretization stencils are used an efficient parallelization of these solvers can be implemented [34, 35].

Computing the average $\overline{U_h^{(k)}}$ by means of the quadrature rule (4), the approximations at the grid nodes $U_h^{(k)}$ are used to determine the solution at the quadrature nodes ξ_j. Denoting the corresponding interpolated values by $I_h\left(U_h^{(k)}\right)(\xi_j)$ one obtains the following quadrature rule:

$$\overline{U_h^{(k)}} = \sum_{j=1}^{nQP} \omega_j \cdot I_h\left(U_h^{(k)}\right)(\xi_j). \tag{12}$$

Finally, stopping criteria for the iteration scheme are required. The first criterion is the maximal number of iterations $Itmax$, while the second one is based on computing the difference of the solutions between two iterations:

$$e_h^{(k+1)} = U_h^{(k+1)} - U_h^{(k)}$$

and the corresponding discrete L^2-norm:

$$\left(\Delta e_h^{(k)}\right)^2 = \left\| e_h^{(k+1)} \right\|_{2,h}^2 = \frac{1}{N}\sum_{i=1}^{N}\left(U_h^{(k+1)}(i) - U_h^{(k)}(i)\right)^2. \tag{13}$$

If $\Delta e_h^{(k)}$ is smaller than a given tolerance TOL, the iteration is stopped.

4 Application of the Zenger Correction

As it has already been mentioned in the previous section, it still remains to derive the regularization of the Dirac measure δ_{x_0}. Thereby, we follow the ideas and concepts presented in [28, 30]. Integrating or differentiating the Delta distribution (in the distributional sense), we obtain one of the functions H_i, $i \in \mathbb{Z}$, which are recursively defined as:

$$H_0(x) = \begin{cases} 0 & \text{for } x \leq 0, \\ 1 & \text{for } x > 0, \end{cases} \quad \text{and} \quad H_i(x) = \begin{cases} \dfrac{d}{dx} H_{i-1}(x) & \text{for } i > 0, \\ \displaystyle\int_{-\infty}^{x} H_{i+1}(\xi)\, d\xi & \text{for } i < 0. \end{cases} \tag{14}$$

Algorithm 3 Iterative solution scheme for the 2D0D coupled problem (3). TOL is a given tolerance and $ITmax$ defines the maximum of the permissible iteration steps

$U_h^{(0)} = \mathbf{0}$
$\overline{U_h^{(0)}} = 0$
Determine the discretization matrix $A_h^{(l)}$ for $l = 5$ (Five point stencil) or $l = 9$ (Nine point stencil) and (9)
Compute the regularization of δ_{x_0} and incorporate the boundary conditions into the vector b_h using (7) for $l = 5$ or (10), (11) for $l = 9$
while $k \leq ITmax$ AND $\Delta e_h^{(k)} > TOL$ **do**
  Compute α_k by means of (8)
  if $l == 5$ **then**
    Update b_h by means of (7)
  else if $l == 9$ **then**
    Update b_h by means of (10)
  end if
  Solve $A_h^{(l)} U_h^{(k+1)} = b_h$
  for $j = 1, \ldots, nQP$ **do**
    Compute ξ_j by (5)
    Interpolate the solution at ξ_j to obtain $I_h\left(U_h^{(k)}\right)(\xi_j)$
    $\overline{U_h^{(k+1)}} += \omega_j \cdot I_h\left(U_h^{(k)}\right)(\xi_j)$
  end for
  Compute $\Delta e_h^{(k)}$ using (13)
  $k = k + 1$
end while

Note that H_0 is the Heaviside step function and H_1 is the Delta distribution. For $i < 0$ it holds that:

$$H_i(x) = \frac{(x)_+^{|i|}}{|i|!}, \qquad (x)_+^{|i|} = \begin{cases} x^{|i|} & \text{for } x \geq 0, \\ 0 & \text{for } x < 0, \end{cases}$$

To define a point source in 2D at $x_0 = \left(x_{0,1}, x_{0,2}\right)^T \in \mathbb{R}^2$, we use a tensor product:

$$H_{x_0}(x) = \prod_{i=1}^{2} H_1\left(x_i - x_{0,i}\right), \quad x = (x_1, x_2)^T \in \mathbb{R}^2$$

For an arbitrary $a \in \mathbb{Z}^2$, $a = (a_1, a_2)^T$ we denote the functions as follows:

$$H_{x_0}^a(x) = \prod_{i=1}^{2} H_{a_i}\left(x_i - x_{0,i}\right).$$

The idea of the Zenger correction is now to integrate $H_i,\ i > 0$ in an analytical way that often such that the result is a sufficiently smooth function $H_i,\ i < 0$. After that we differentiate this function numerically as often as we integrated the right hand side. If the source term of the PDE is integrated and differentiated n times, the result is defined as **Zenger correction of order n**.

Modeling a point source term in 2D at $x_0 = \left(x_{0,1}, x_{0,2}\right)^T \in \mathbb{R}^2$, i.e., $H_{x_0}^{(1,1)}$, one has to integrate two times in each space dimensions, which yields the function $H_{x_0}^{(-1,-1)}$. This function is two times differentiated concerning the two space dimensions. Denoting the numerical derivative by

$$\delta_x^{(2,2)} = \delta_{x_1}^2 \circ \delta_{x_2}^2, \quad x = (x_1, x_2)^T$$

we have the following numerical equivalent:

$$\delta_x^{(2,2)} H_{x_0}^{(-1,-1)}(x) = \prod_{i=1}^{2} \delta_{x_i}^2 H_{-1}\left(x_i - x_{0,i}\right), \tag{15}$$

where $\delta_{x_i}^2$ is the second numerical derivative with respect to x_i. For an even $n \in \mathbb{N}$ and a mesh size h the numerical derivative of $H_{1-n}(x_i)$ with $1 - n < 0$ and n even: in x_i is given by Köstler and Rüde [28]:

$$\delta_{x_i}^n H_{1-n}(x_i) = \begin{cases} 0, & \text{if } |x_i| \geq \frac{n}{2}h, \\ \sum_{j=0}^{n} \frac{(-1)^j}{h^n} \binom{n}{j} H_{1-n}\left(x_i + \left(j - \frac{n}{2}\right)h\right), & \text{if } |x_i| < \frac{n}{2}h. \end{cases}$$

On closer examination, one observes that the above formula has the shape of a B-spline. Consequently we can use for the computation and evaluation of $\delta_{x_i}^n H_{1-n}\left(x_i - x_{0,i}\right)$ all the efficient and robust algorithms which are known in context with splines [36]. For $n = 2$ one obtains the following numerical equivalent corresponding to the Zenger correction of second order:

$$\delta_{x_i}^2 H_{-1}\left(x_i - x_{0,i}\right) = \begin{cases} 0, & \text{if } \left|x_i - x_{0,i}\right| \geq h, \\ \frac{h - \left|x_i - x_{0,i}\right|}{h^2}, & \text{if } \left|x_i - x_{0,i}\right| < h. \end{cases}$$

Combining a Zenger correction of order 2 with a discretization stencil of order 2 (see Table 1), one can prove that the resulting numerical scheme for a problem like (2) converges locally with $O\left(h^2\right)$ [28]. In order to design numerical schemes with higher order, we require a right hand side that is smoother. As it can be seen in Fig. 1 (left), the graph of a second order Zenger correction is continuous but not

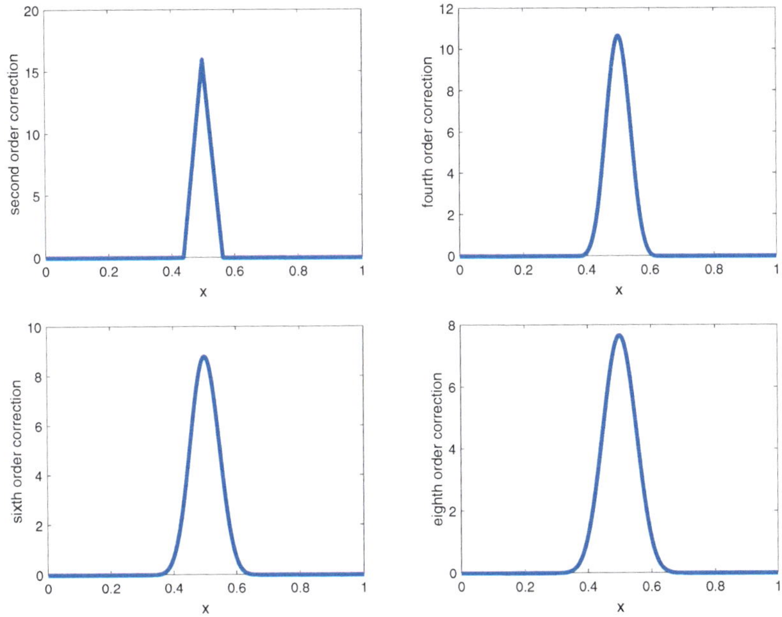

Fig. 1 Zenger corrections of order $n \in \{2, 4, 6, 8\}$ for $h = 1/16$ and $x_{0,1} = 0.5$

differentiable. For an approximation order of higher order, the right hand side of the PDE has to be smoother. To obtain a smoother right hand side a Zenger correction of higher order has to be computed. A Zenger correction of fourth order is given by:

$$\delta_x^{(4,4)} H_{x_0}^{(-3,-3)}(x) = \prod_{i=1}^{2} \delta_{x_i}^4 H_{-3}\left(x_i - x_{0,i}\right). \tag{16}$$

On the right hand side of Fig. 1 the graph of a fourth order Zenger correction in a single dimension is depicted. Obviously the fourth order Zenger correction is a smooth function. Locally is given by a third order polynomial. All in all, it can be shown that the fourth order Zenger correction is twice continuous differentiable. This procedure can be further extended resulting in even smoother approximations of the Dirac measure with larger support and decreasing maximum.

5 Numerical Tests

In this section, we want to investigate the convergence behavior of the numerical method that has been described in the previous two sections. Thereby, three different issues are considered. The first one is studying whether the Zenger Correction induces no pollution effect, i.e., whether the numerical method converges locally with optimal order. For this purpose an error norm is considered excluding a ball of radius R around the singularity. It is given by:

$$e_R = \frac{\left(\frac{1}{|I_R|}\sum_{(i,j)\in I_R} (u_h - u_e)^2 \left(x_{ij}\right)\right)^{\frac{1}{2}}}{\left(\frac{1}{N}\sum_{i,j=1}^{N_d} u_e^2 \left(x_{ij}\right)\right)^{\frac{1}{2}}}.$$

The index set I_R is defined as:

$$I_R = \left\{(i, j) \in \{1, \ldots, N_d\} \times \{1, \ldots, N_d\} \,\middle|\, x_{ij} \in [0, 1]^2 \setminus B_R\left(x_0\right)\right\}.$$

This region is of great interest for the application area of this paper, since within $B_R\left(x_0\right)$ the solution is governed by the blood flow model. In order to test the convergence of the different approximation schemes, the boundary conditions u_B are chosen such that

$$u_e\left(x\right) = -U\frac{RL_p}{1 - RL_p \ln R} \ln r\left(x\right) \tag{17}$$

is the exact solution of (2). Thereby, $r\left(x\right)$ is defined as follows:

$$r\left(x\right) = \sqrt{\left(x_1 - x_{0,1}\right)^2 + \left(x_2 - x_{0,2}\right)^2}.$$

A numerical approximation of u_e is depicted in Fig. 2. Another interesting issue is concerned with the computation of the exchange term E in the source term of (2), which is given by:

$$E = 2\pi RL_p\left(U - \overline{u}\right).$$

E is of great relevance in terms of simulating blood flow and transport processes, since it quantifies the amount of blood volume transferred from the vascular system to the tissue matrix or vice versa. Considering a diffusion dominated transport process within the tissue matrix a precise approximation of E is even more important than an accurate approximation of the pressure in each point of the tissue domain. However, on coarse meshes and for small radii R the average $\overline{u}$ can not be computed in an accurate way. It may even happen that the circle $\partial B_R\left(x_0\right)$ on which u is averaged, is completely contained in the support of the Zenger correction. Therefore, a strategy is required exploiting the approximation properties of the Zenger correction at a distance of the singularity to compute the average $\overline{u}$. This allows us to compute E with a low approximation error.

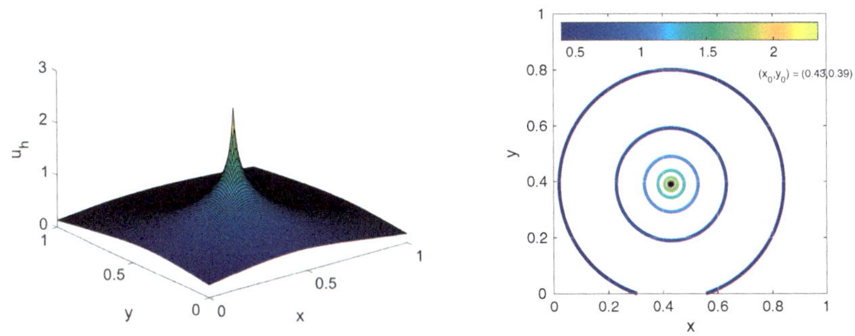

Fig. 2 Left: numerical approximation U_h of the exact solution u_e (17) for $R = 0.2$, $U = 3.0$, $L_p = 1.0$, $x_0 = (0.43, 0.39)$ and $h = 1/128$. Right: contour plot of the solution u_e

Finally, the fixed point iteration described in Algorithm 3 is investigated. In particular, it is tested in a numerical way, under which conditions, the fixed point iteration is converging. Thereby, the model parameters R, L_p and U are varied. For all the upcoming experiments, the following mesh sizes are considered:

$$h = \frac{1}{8 \cdot 2^l}, \quad l \in \{0, 1, 2, 3, 4, 5\}.$$

The maximal iteration number $Itmax$ in Algorithm 3, is set to 20 and the error tolerance TOL is given by: $TOL = \min\left(1.0 \cdot 10^{-8}, 0.1 \cdot h^2\right)$ in case of the Five point stencil and $TOL = \min\left(1.0 \cdot 10^{-8}, 0.1 \cdot h^4\right)$ in case of the Nine point stencil to ensure that the updates of the iteration scheme are in the range of the expected approximation error or below. In case of the experiments that are presented in Sects. 5.1 and 5.2 Algorithm 3 converges latest after 14 iterations. For the quadrature rule (4), we choose: $nGP = 1024$ and $\omega_j = 1/nGP$. In order to keep the operations in Algorithm 3 local, we use a bilinear interpolation to compute the values $I_h\left(U_h^{(k)}\right)(\xi_j)$, i.e., all the operations in Algorithm 3 can be implemented in a stencil based way.

5.1 Local Convergence Behavior

For this first test, the model parameters are chosen as follows:

$$R \in \{0.01,\ 0.05,\ 0.10,\ 0.15\},\ L_p = 1.0 \text{ and } U = 3.0.$$

Table 2 Local convergence behavior for a Five point stencil, a Zenger correction of second order and $x_0 = (0.50, 0.50)$. The abbreviation *e.c.o* stands for *estimated convergence order*

	$R = 0.15$		$R = 0.1$		$R = 0.05$		$R = 0.01$	
l	e_R	*e.c.o.*	e_R	*e.c.o.*	e_R	*e.c.o.*	e_R	*e.c.o.*
0	1.10*e*–02	–	1.46*e*–02	–	1.50*e*–02	–	1.66*e*–02	–
1	2.34*e*–03	2.23	3.87*e*–03	1.91	7.31*e*–03	1.03	8.21*e*–03	1.02
2	6.02*e*–04	1.96	8.34*e*–04	2.21	1.94*e*–03	1.91	3.79*e*–03	1.12
3	1.41*e*–04	2.09	2.19*e*–04	1.93	4.21*e*–04	2.20	1.84*e*–03	1.04
4	3.50*e*–05	2.01	5.43*e*–05	2.01	1.10*e*–04	1.93	6.84*e*–04	1.43
5	8.76*e*–06	2.00	1.34*e*–05	2.02	2.74*e*–05	2.01	1.49*e*–04	2.20

Moreover, two different points x_0 are considered:

$$x_0 \in \{(0.5, 0.5), (0.43, 0.39)\}.$$

In case of the first point the Dirac measure is located at a grid node of the computational mesh, while the second point does not coincide with a grid node. The Five point stencil is combined with a second order Zenger correction. As it has been shown in [29, Theorem 4.1] this numerical scheme should converge locally with second order. Using the Nine point stencil, a fourth order Zenger correction is considered as in [37]. According to Table 6 of this work a local fourth order convergence can be observed for a Poisson problem with a point source term. In Table 2 the results for $x_0 = (0.5, 0.5)$ and the second order discretization are listed. For larger radii $R = 0.15$ and $R = 0.10$ an average convergence rate of 2 can be observed. Smaller radii show a pre-asymptotic range. This is due to the fact that the average $\overline{u}$ is not well approximated on coarse meshes. Considering the convergence rates for $R = 0.05$ and $l \in \{2, \dots, 5\}$, the average convergence rate is given by:

$$\frac{1}{4}(1.91 + 2.20 + 1.93 + 2.01) = 2.01,$$

while for the smallest radius an optimal convergence rate of 2 is obtained only for the last refinement step from $l = 4$ to $l = 5$. Similar observations can be made for $x_0 = (0.43, 0.39)$ (see Table 3). However, for some radii the pre-asymptotic ranges are extended. The method shows for $R = 0.1$ an average convergence rate of 2 regarding the refinement levels $l \in \{2, \dots, 5\}$. $R = 0.05$ exhibits an average convergence rate of 2.32 for the last two refinements. Tables 4 and 5 show the results for the fourth order Zenger correction. Obviously, the approximation errors are remarkably reduced compared to the second order Zenger correction. For $R = 0.01$ the errors at the refinement level $l = 5$ are a factor 10 smaller.

Contrary to the second order Zenger correction the convergence rates are not stable and the average convergence rates are not equal to the optimal convergence order of 4. The reason why the convergence rates are not optimal, is due to the fact that a low order interpolation method is used in terms of the quadrature rule.

Table 3 Local convergence behavior for a Five point stencil, a Zenger correction of second order and $x_0 = (0.43, 0.39)$. The abbreviation *e.c.o* stands for *estimated convergence order*

	$R = 0.15$		$R = 0.1$		$R = 0.05$		$R = 0.01$	
l	e_R	*e.c.o.*	e_R	*e.c.o.*	e_R	*e.c.o.*	e_R	*e.c.o.*
0	1.21*e*–02	–	1.29*e*–02	–	1.86*e*–02	–	1.85*e*–02	–
1	2.45*e*–03	2.30	4.11*e*–03	1.65	5.95*e*–03	1.65	1.40*e*–02	0.40
2	5.61*e*–04	2.13	9.08*e*–04	2.18	1.45*e*–03	2.03	4.21*e*–03	1.73
3	2.17*e*–04	1.37	3.54*e*–04	1.36	7.32*e*–04	0.99	2.38*e*–03	0.82
4	3.54*e*–05	2.62	5.54*e*–05	2.68	1.11*e*–04	2.72	6.95*e*–04	1.78
5	9.25*e*–06	1.94	1.43*e*–05	1.95	2.92*e*–05	1.92	1.58*e*–04	2.14

Moreover, as in case of the second order Zenger correction, the average $\overline{u}$ is not approximated with sufficient precision. A way how the convergence rates for the fourth order scheme can be improved is discussed in the next section.

5.2 Accurate Computation of the Exchange Term

In order to compute the exchange term

$$E = 2\pi L_p R \left(U - \overline{u}\right),$$

modeling the flux between the vascular system and tissue, it is crucial to approximate the average $\overline{u}$ with sufficient accuracy. However, if the radius R is small compared to the mesh size h and the support of the Zenger correction, one is faced with the problem that the average value $\overline{U_h^{(k)}}$ is polluted by the poor approximation of the solution in the vicinity of the singularity. This might deteriorate not only the convergence with respect to $\overline{U_h^{(k)}}$, but also the convergence with respect to $U_h^{(k)}$ as it can be seen in Table 4 and 5. To get rid of this drawback, a technique is proposed, which uses average values on larger circles around the singularity to compute a more accurate average value $\overline{u}$. Observing that the numerical error of a Zenger correction scheme does not pollute the solution at a larger distance to the singularity, one can conclude that the coupling values $\overline{U_h^{(k)}}_{R_e}$ on a circle $\partial B_{R_e}(x_0)$, $R_e > R$ can be computed more accurately. In the following, it is outlined, how this value can be used to compute $\overline{U_h^{(k)}}$ without pollution. The solution in the k-th iteration of Algorithm 3 can be approximated by a sum of the fundamental solution

$$u_F^{(k+1)} = -L_p R \cdot \left(U - \overline{u^{(k)}}\right) \cdot \ln r$$

Table 4 Local convergence behavior for a Nine point stencil, a Zenger correction of fourth order and $x_0 = (0.50, 0.50)$. The abbreviation *e.c.o* stands for *estimated convergence order*

	$R = 0.15$		$R = 0.10$		$R = 0.05$		$R = 0.01$	
l	e_R	*e.c.o.*	e_R	*e.c.o.*	e_R	*e.c.o.*	e_R	*e.c.o.*
0	2.87*e*–03	–	1.57*e*–02	–	2.03*e*–02	–	2.01*e*–02	–
1	1.32*e*–04	4.44	1.06*e*–03	3.89	7.43*e*–03	1.45	1.03*e*–02	0.97
2	2.45*e*–05	2.43	6.66*e*–05	3.99	5.99*e*–04	3.63	4.78*e*–03	1.10
3	5.06*e*–06	2.27	2.65*e*–06	4.65	3.79*e*–05	3.98	1.89*e*–03	1.34
4	8.70*e*–07	2.54	1.43*e*–06	0.89	1.54*e*–06	4.62	1.94*e*–04	3.29
5	3.97*e*–08	4.46	3.66*e*–07	1.96	7.89*e*–07	0.96	1.36*e*–05	3.84

Table 5 Local convergence behavior for a Nine point stencil, a Zenger correction of fourth order and $x_0 = (0.43, 0.39)$. The abbreviation *e.c.o* stands for *estimated convergence order*

	$R = 0.15$		$R = 0.10$		$R = 0.05$		$R = 0.01$	
l	e_R	*e.c.o.*	e_R	*e.c.o.*	e_R	*e.c.o.*	e_R	*e.c.o.*
0	3.88*e*–03	–	1.08*e*–02	–	1.90*e*–02	–	1.88*e*–02	–
1	1.86*e*–04	4.38	8.49*e*–04	3.67	5.61*e*–03	1.76	2.22*e*–02	–0.24
2	1.49*e*–05	3.64	5.53*e*–05	3.94	2.68*e*–04	4.39	5.39*e*–03	2.04
3	1.08*e*–06	3.80	2.73*e*–06	4.34	1.90*e*–05	3.82	1.38*e*–03	1.97
4	7.68*e*–07	0.48	1.21*e*–06	1.18	1.27*e*–06	3.91	2.10*e*–04	2.72
5	2.62*e*–08	4.88	2.82*e*–07	2.10	4.68*e*–07	1.44	1.27*e*–05	4.05

and a smooth function $u_S^{(k+1)}$ solving the following Laplace equation:

$$-\Delta u_S^{(k+1)} = 0 \text{ in } (0,1)^2,$$
$$u_S^{(k+1)} = u_e - u_F^{(k+1)} \text{ on } \partial (0,1)^2.$$

Having this notation at hand, we can approximate the average values on $B_R(x_0)$ and $B_{R_e}(x_0)$ by:

$$\overline{u^{(k+1)}} \approx \overline{u_F^{(k+1)}} + \overline{u_S^{(k+1)}} = -L_p R \cdot \left(U - \overline{u^{(k)}}\right) \cdot \ln R + \overline{u_S^{(k+1)}},$$

$$\overline{u^{(k+1)}}_{R_e} \approx \overline{u_F^{(k+1)}}_{R_e} + \overline{u_S^{(k+1)}}_{R_e} = -L_p R \cdot \left(U - \overline{u^{(k)}}\right) \cdot \ln R_e + \overline{u_S^{(k+1)}}_{R_e}.$$

Since the function $u_S^{(k+1)}$ exhibits the mean value property [38]:

$$u_S^{(k+1)}(x_0) = \overline{u_S^{(k+1)}}_{R_e} = \overline{u_S^{(k+1)}},$$

we have:

$$\overline{u^{(k+1)}} \approx \overline{u^{(k+1)}}_{R_e} + L_p R \cdot \left(U - \overline{u^{(k)}}\right) \ln\left(\frac{R_e}{R}\right). \tag{18}$$

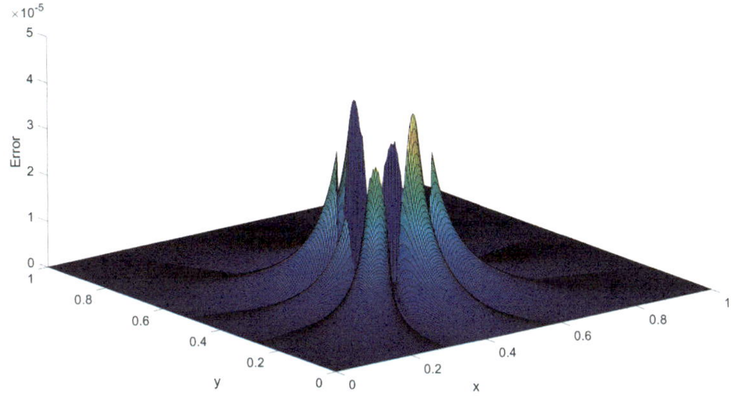

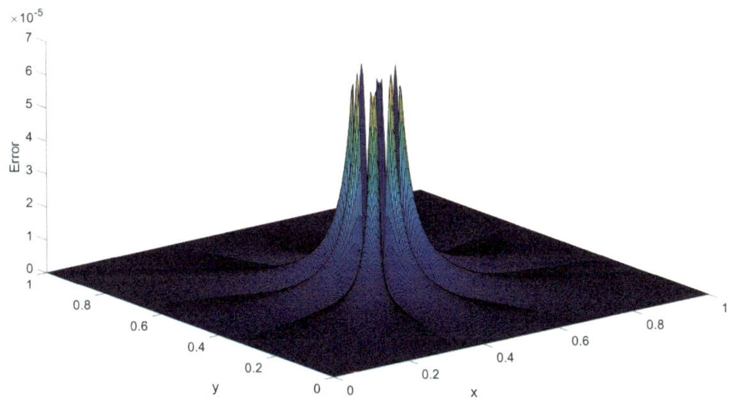

Fig. 3 Top: approximation errors in $(0,1)^2 \setminus B_R(x_0)$ for $R = 0.1$ and $x_0 = (0.43, 0.39)$. Bottom: approximation errors in $(0,1)^2 \setminus B_R(x_0)$ for $R = 0.05$ and $x_0 = (0.50, 0.50)$. In both cases the error is set to zero in $B_R(x_0)$. All the other parameters are taken over from Sect. 5.1. For the numerical solution a Five point stencil combined with a second order Zenger correction is used. The grid width h is given by $1/256$

In the following, we refer to (18) as extrapolation of the average, since we use an average computed outside of a region in which the solution is not well approximated. An interesting issue is now how the extrapolation radius R_e should be chosen. It can be observed in Fig. 3 that the numerical error decreases with increasing distance to the singularity. Thus R_e should be chosen as large as possible. This means that the circle $\partial B_{R_e}(x_0)$ should be contained in the computational domain. Moreover, it was assumed for the derivation of (18) that the smooth component of the solution is harmonic. However, this is not the case, if $B_{R_e}(x_0)$ contains a further singularity.

Table 6 Local convergence behavior for a Nine point stencil, a Zenger correction of fourth order and $x_0 = (0.50, 0.50)$. Contrary to Table 4, the extrapolation technique for $\overline{u}$ has been applied. The abbreviation *e.c.o* stands for *estimated convergence order*

	$R = 0.15$		$R = 0.1$		$R = 0.05$		$R = 0.01$	
l	e_R	$e.c.o.$	e_R	$e.c.o.$	e_R	$e.c.o.$	e_R	$e.c.o.$
0	3.11*e*–03	–	1.35*e*–02	–	1.35*e*–02	–	1.35*e*–02	–
1	1.33*e*–04	4.54	3.07*e*–04	5.46	6.52*e*–03	1.05	6.52*e*–03	1.04
2	1.00*e*–05	3.73	2.42*e*–05	3.66	1.51*e*–04	5.43	3.23*e*–03	1.01
3	6.42*e*–07	3.97	1.94*e*–06	3.64	1.19*e*–05	3.67	1.61*e*–03	1.00
4	3.64*e*–08	4.14	1.20*e*–07	4.02	9.59*e*–07	3.63	1.80*e*–04	3.16
5	3.30*e*–09	3.46	7.79*e*–09	3.94	5.92*e*–08	4.02	7.91*e*–06	4.51

Table 7 Local convergence behavior for a Nine point stencil, a Zenger correction of fourth order and $x_0 = (0.43, 0.39)$. Contrary to Table 5, the extrapolation technique for $\overline{u}$ has been applied. The abbreviation *e.c.o* stands for *estimated convergence order*

	$R = 0.15$		$R = 0.1$		$R = 0.05$		$R = 0.01$	
l	e_R	$e.c.o.$	e_R	$e.c.o.$	e_R	$e.c.o.$	e_R	$e.c.o.$
0	4.37*e*–03	–	1.02*e*–02	–	1.21*e*–02	–	1.21*e*–02	–
1	1.86*e*–04	4.55	4.64*e*–04	4.46	5.14*e*–03	1.24	2.22*e*–02	−0.87
2	1.21*e*–05	3.95	4.64*e*–05	3.32	2.48*e*–04	4.37	3.25*e*–03	2.77
3	7.33*e*–07	4.04	2.49*e*–06	4.22	1.90*e*–05	3.70	1.17*e*–03	1.47
4	4.25*e*–08	4.11	1.24*e*–07	4.32	9.23*e*–07	4.37	1.99*e*–04	2.56
5	2.77*e*–09	3.94	8.28*e*–09	3.91	6.61*e*–08	3.80	9.42*e*–06	4.40

Thus, if several point sources have to be considered, one has to make sure that the extrapolation circles assigned to the different point sources do not include other point sources. Since in this work only a single point source is considered, R_e is chosen equal to the minimal distance of x_0 to the boundary of the unit square:

$$R_e = \min_{x \in \partial(0,1)^2} \|x - x_0\|_2 \,.$$

Using the extrapolation technique, the numerical tests from Sect. 5.1 are repeated with respect to the fourth order Zenger correction and the Nine point stencil. This means that updating α_{k+1} in Algorithm 3, (18) is used to compute $\overline{U_h^{(k+1)}}$. The corresponding results are contained in Tables 6 and 7. Contrary to the corresponding results in Tables 4 and 5 an average convergence rate of 4 is obtained for the larger radii $R = 0.15$ and $R = 0.10$. In case of the smaller radii $R = 0.05$ and $R = 0.01$ there is a pre-asymptotic range. For $R = 0.05$, an optimal average convergence order is observed for the levels $l \in \{2, 3, 4, 5\}$, while for $R = 0.01$ the optimal convergence order 4 occurs in the last refinement step. This means that the extrapolation rule yields optimal average convergence rates for the fourth order Zenger correction. Using the standard computation rule for the average term deteriorates the convergence behavior of the fourth order solution

Table 8 Error $\overline{e}_R$ for a Five point stencil, a Zenger correction of second order and $x_0 = (0.50, 0.50)$. The term (no ex.) indicates that the extrapolation rule (18) is not applied, while the term (ex.) indicates that the extrapolation rule (18) is applied

	$R = 0.1$		$R = 0.05$		$R = 0.01$	
l	$\overline{e}_R$ (no ex.)	$\overline{e}_R$ (ex.)	$\overline{e}_R$ (no ex.)	$\overline{e}_R$ (ex.)	$\overline{e}_R$ (no ex.)	$\overline{e}_R$ (ex.)
0	5.16e–03	4.69e–04	5.00e–03	2.61e–04	2.98e–02	6.17e–05
1	1.57e–03	1.06e–04	3.03e–03	5.87e–05	1.46e–02	1.40e–05
2	3.27e–04	2.41e–05	9.09e–04	1.34e–05	3.46e–03	3.20e–06
3	6.51e–06	5.51e–06	1.75e–04	3.11e–06	1.09e–03	7.44e–07
4	6.15e–06	1.31e–06	5.47e–06	7.76e–07	1.99e–04	1.76e–07
5	1.43e–06	3.55e–07	3.93e–06	1.91e–07	6.39e–05	4.35e–08

Table 9 Error $\overline{e}_R$ for a Nine point stencil, a Zenger correction of fourth order and $x_0 = (0.50, 0.50)$. The term (no ex.) indicates that the extrapolation rule (18) is not applied, while the term (ex.) indicates that the extrapolation rule (18) is applied

	$R = 0.1$		$R = 0.05$		$R = 0.01$	
l	$\overline{e}_R$ (no ex.)	$\overline{e}_R$ (ex.)	$\overline{e}_R$ (no ex.)	$\overline{e}_R$ (ex.)	$\overline{e}_R$ (no ex.)	$\overline{e}_R$ (ex.)
0	1.35e–02	1.74e–04	4.17e–02	9.73e–05	4.47e–02	2.30e–05
1	4.91e–03	3.32e–05	7.54e–03	1.86e–05	2.78e–02	4.39e–06
2	3.14e–04	6.09e–06	2.75e–03	3.40e–06	1.35e–02	8.05e–07
3	9.99e–06	1.11e–06	1.75e–04	6.19e–07	3.66e–03	1.46e–07
4	7.18e–06	2.14e–07	5.58e–06	1.20e–07	4.40e–04	2.83e–08
5	1.72e–06	4.18e–08	4.01e–06	2.34e–08	5.59e–05	5.53e–09

scheme. Consequently, the approximation errors are lower if the extrapolation rule is considered.

Next, we turn our attention again to the issue raised at the beginning of this subsection, i.e., the computation of the exchange term E. As it has already been mentioned the approximation error of E depends on the approximation of $\overline{u}$. Therefore, the following error term is studied:

$$\overline{e}_R = \left|\overline{u_e} - \overline{U_h}\right|,$$

where $\overline{U_h}$ is obtained from Algorithm 3 after the last iteration is finished. In Tables 8 and 9 the errors $\overline{e}_R$, $R \in \{0.01, 0.05, 0.1\}$ are listed for the second order and fourth order Zenger correction. The point source is concentrated at $x_0 = (0.50, 0.50)$. For each radius, we compare the convergence of $\overline{e}_R$ for two different cases: In the first case the average of the solution is computed by means of (12), while in the second case the extrapolation rule (18) is considered. As it can be seen in the upcoming tables, the errors for the average are remarkably reduced by applying the extrapolation rule (18). In particular on coarse grids and for $R \in \{0.01, 0.05\}$, the errors can be reduced by three orders of magnitude. Moreover, we observe that the fourth order correction scheme yields lower errors for $l \in \{2, 3, 4, 5\}$ compared to the second order correction scheme, if the extrapolation rule (18) is used.

These results imply that the exchange term E can be computed with high precision even on coarse meshes. If one is only interested in determining the fluid mass transferred between the vascular system and tissue, it is much more important to have precise data with respect to E, whereas the convergence of the local error e_R is less important.

5.3 Convergence of the Iterative Solution Scheme

Studying the convergence behavior of the iteration scheme described in Algorithm 3, we try to motivate a possible convergence criterion. This criterion, will help us to rate the data resulting from varying the different model parameter. Deriving this criterion, we recall some ideas from [19, Section 3.2.2]. A closer examination shows that the first sub step in each iteration is the discrete version of the following Dirac problem:

$$-\Delta u^{(k+1)} = 2\pi R L_p \left(U - \overline{u^{(k)}} \right) \delta_{x_0}. \tag{19}$$

From literature we know that the solution of (19) in the sense of distributions is given by Evans [38]:

$$u^{(k+1)} = -RL_p \left(U - \overline{u^{(k)}} \right) \ln r. \tag{20}$$

For the first three iterations, we obtain:

$$\begin{aligned}
u^{(1)} &= -RL_p \left(U - \overline{u^{(0)}} \right) \ln r = -RL_p U \ln r \\
u^{(2)} &= -RL_p \left(U - \overline{u^{(1)}} \right) \ln r = -RL_p U \left(1 + RL_p \ln R \right) \ln r \\
u^{(3)} &= -RL_p \left(U - \overline{u^{(2)}} \right) \ln r \\
&= -RL_p U \left(1 + \left(RL_p \ln R \right) + \left(RL_p \ln R \right)^2 \right) \ln r
\end{aligned}$$

By induction with respect to k one can prove:

$$u^{(k+1)} = -RL_p U \sum_{l=0}^{k} \left(RL_p \ln R \right)^l \ln r. \tag{21}$$

From this it follows:

$$\lim_{k\to\infty} u^{(k+1)} = -RL_pU \sum_{l=0}^{\infty} \left(RL_p \ln R\right)^l \ln r$$
$$= -U \frac{RL_p}{1 - RL_p \ln R} \ln r = u_e,$$

where the series from above is convergent, if

$$\left|RL_p \ln R\right| < 1 \tag{22}$$

holds. A closer look at inequality (22) suggests that the iteration is diverging, if the radius R and the permeability L_p are sufficiently large. Moreover, it seems that the choice of U does not influence the convergence of the fixed point iteration. Based on these observations, we choose the radius from the following set:

$$R \in \{0.0375,\ 0.075,\ 0.15,\ 0.30\}.$$

Concerning the iteration scheme only the parameter $Itmax$ is changed. Here, we make the following choice: $Itmax = 30$. Using these parameters, two test series with the following parameters are performed:

- Test (1):

$$U = 1,\ L_p \in \{3.0,\ 4.0,\ 5.0,\ 6.0\}.$$

- Test (2):

$$L_p = 1,\ U \in \{1.0,\ 2.0,\ 3.0,\ 4.0\}.$$

For the numerical solution $h = 1/256$, a Five point stencil and a Zenger correction of second order without extrapolation is used. In order to illustrate the convergence behavior of the fixed point iteration, update (13) is reported for each iteration k. The results for Test (1) are provided in Fig. 4, while the results for Test (2) can be seen in Fig. 5. Considering the different results, the qualitative behavior that can be motivated by (22) is confirmed. In Fig. 4 one can observe that the iteration is diverging or slowly converging, if L_p and R are relatively large. In case of the smaller radii and smaller permeabilities the iteration is converging faster, in particular, if (22) is fulfilled. However, (22) is not a sharp condition. For $L_p = 5$ and $R = 0.3$, we obtain:

$$\left|RL_p \ln R\right| \approx 1.81 > 1,$$

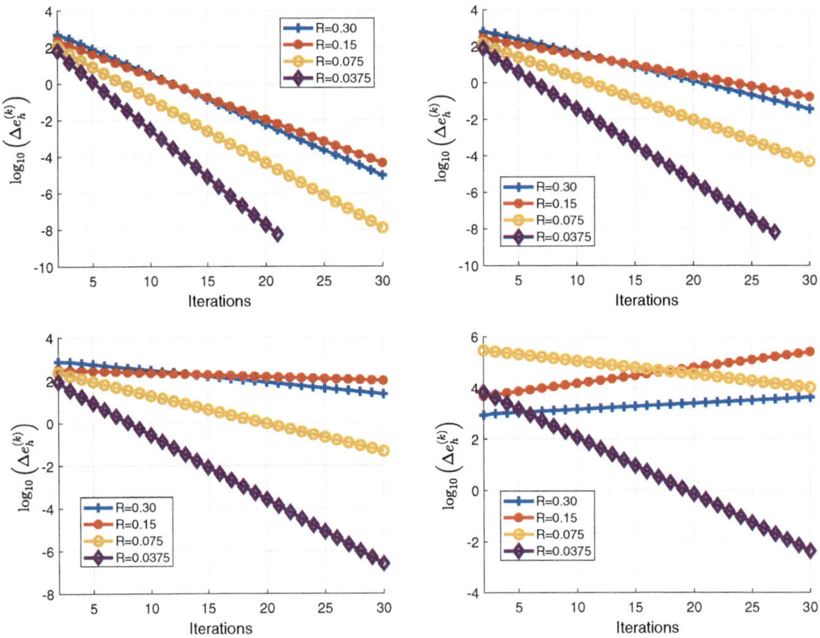

Fig. 4 The four sub figures show the updates $\Delta e_h^{(k)}$ for each iteration k. Each sub figure contains four graphs corresponding to the radii $R \in \{0.0375, 0.075, 0.15, 0.30\}$. For the sub figure top left $L_p = 3.0$ is used, while for the sub figure top right $L_p = 4.0$ is considered. At the bottom $L_p = 5.0$ (left) and $L_p = 6.0$ (right) is chosen. In order to be able to distinguish the different curves of the updates, the logarithm with respect to 10 is shown

which means that in this case the iteration scheme should be diverging. Considering the bottom left sub figure in Fig. 4, one can see that the corresponding updates are slowly decreasing, but not increasing as it would have been expected. The reason for this behavior might be due to the fact that the derivation of (22) is based on a sequence of continuous problems and not on the discrete problems in Algorithm 3. To derive a sharp stability condition, a sequence of discrete problems given by (6) has to be analyzed. Contrary to L_p, U has almost no influence on the convergence behavior of the fixed point iteration. As the graphs in Fig. 5 suggest, the graphs in the different sub figures are quite similar. This is in accordance with (22), which is independent of U.

6 Conclusions and Outlook

In this work, we have studied an elliptic Dirac problem that is closely related to a 3D-1D coupled problem describing flow in microvascular networks. A well-known difficulty associated with Dirac problems is due to the fact that solutions

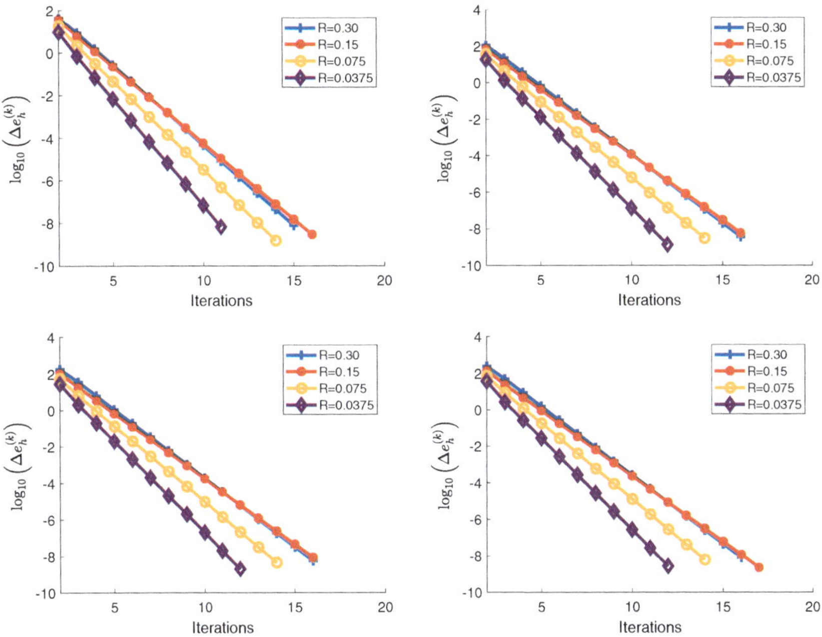

Fig. 5 The four sub figures show the updates $\Delta e_h^{(k)}$ for each iteration k. Each sub figure contains four graphs corresponding to the radii $R \in \{0.0375,\ 0.075,\ 0.15,\ 0.30\}$. For the sub figure top left $U = 1.0$ is used, while for the sub figure top right $U = 2.0$ is considered. At the bottom $U = 3.0$ (left) and $U = 4.0$ (right) is chosen. In order to be able to distinguish the different curves of the updates, the logarithm with respect to 10 is shown

become singular at positions where the Dirac source term is concentrated. In order to handle this problem in terms of numerical solvers, a Finite Difference method combined with a regularization of the Dirac source is used. In literature this type of regularization is known as Zenger correction. By means of this regularization, it is possible to design numerical solution methods converging locally with optimal order. Thereby, only stencil based operations are used, allowing a matrix-free implementation of the algorithm, the application of efficient multigrid methods and parallelization. Since an average of the solution is occurring in the source term of the elliptic Dirac problem, we apply a fixed point iteration such that in each iteration a standard Dirac problem has to be solved. The performance of this algorithm is demonstrated by means of three different scenarios.

Based on the results presented in this work, many interesting issues can be investigated. One of them is to apply the concepts of this work to 3D-1D coupled problems. In this context, it has to be clarified how Dirac source terms concentrated on lines can be handled by means of regularizations. Moreover the coupling with the 1D problem has to be added. After testing this algorithm by means of simple problems with an analytical solution, the simulation of blood flow in a complex network supplying a tissue block can be addressed. Apart form that, it would be interesting to study regularizations adapted to other discretization methods such as Finite element methods. Finally, the convergence of the iterative coupling scheme should be studied also in a mathematical precise way both for the model problem discussed in this paper and the 3D-1D coupled problem. Maybe suitable terms can be derived and added to this scheme such that its speed of convergence is accelerated.

References

1. Chapman, S.J., Shipley, R.J., Jawad, R.: Multiscale modeling of fluid transport in tumors. Bullet. Math. Biol. **70**, 2334–2357 (2008)
2. Shipley, R.J., Chapman, S.J.: Multiscale modelling of fluid and drug transport in vascular tumours. Bullet. Math. Biol. **72**, 1464–1491 (2010)
3. Kremheller, J., Vuong, A.T., Schrefler, B.A., Wall, W.A.: An approach for vascular tumor growth based on a hybrid embedded/homogenized treatment of the vasculature within a multiphase porous medium model. Int. J. Numer. Methods Biomed. Eng. **35**(11), e3253 (2019)
4. Fritz, M., Jha, P.K., Köppl, T., Oden, J.T., Wohlmuth, B.: Analysis of a new multispecies tumor growth model coupling 3D phase-fields with a 1D vascular network. Nonlinear Anal. Real World Appl. **61**, 103331 (2021)
5. Nabil, M., Decuzzi, P., Zunino, P.: Modelling mass and heat transfer in nano-based cancer hyperthermia. R. Soc. Open Sci. **2**(10), 150447 (2015)
6. Gould, I.G., Tsai, P., Kleinfeld, D., Linninger, A.: The capillary bed offers the largest hemodynamic resistance to the cortical blood supply. J. Cerebral Blood Flow Metabolism **37**(1), 52–68 (2017)
7. Köppl, T., Vidotto, E., Wohlmuth, B.: A 3D–1D coupled blood flow and oxygen transport model to generate microvascular networks. Int. J. Numer. Methods Biomed. Eng. **36**(10), e3386 (2020)
8. Schneider, M., Reichold, J., Weber, B., Székely, G., Hirsch, S.: Tissue metabolism driven arterial tree generation. Med. Image Anal. **16**(7), 1397–1414 (2012)
9. Smith, A.F., Doyeux, V., Berg, M., Peyrounette, M., Haft-Javaherian, M., Larue, A.E., Lorthois, S.: Brain capillary networks across species: a few simple organizational requirements are sufficient to reproduce both structure and function. Front. Physiol. **10**, 233 (2019)
10. Secomb, T.W., Alberding, J.P., Hsu, R., Dewhirst, M.W., Pries, A.R.: Angiogenesis: an adaptive dynamic biological patterning problem. PLoS Comput. Biol. **9**(3), e1002983 (2013)
11. Ehlers, W., Wagner, A.: Multi-component modelling of human brain tissue: a contribution to the constitutive and computational description of deformation, flow and diffusion processes with application to the invasive drug-delivery problem. Comput. Methods Biomech. Biomed. Eng. **18**(8), 861–879 (2015)
12. Vidotto, E., Koch, T., Köppl, T., Helmig, R., Wohlmuth, B.: Hybrid models for simulating blood flow in microvascular networks. Multiscale Model. Simul. **17**(3), 1076–1102 (2019)
13. D'Angelo, C., Quarteroni, A.: On the coupling of 1d and 3d diffusion-reaction equations: application to tissue perfusion problems. Math. Models Methods Appl. Sci. **18**(08), 1481–1504 (2008)
14. Reichold, J., Stampanoni, M., Keller, A.L., Buck, A., Jenny, P., Weber, B.: Vascular graph model to simulate the cerebral blood flow in realistic vascular networks. J. Cerebral Blood Flow Metabolism **29**(8), 1429–1443 (2009)
15. D'Angelo, C.: Multiscale modelling of metabolism and transport phenomena in living tissues. EPFL (2007)

16. Gong, W., Wang, G., Yan, N.: Approximations of elliptic optimal control problems with controls acting on a lower dimensional manifold. SIAM J. Control Optim. **52**(3), 2008–2035 (2014)
17. D'Angelo, C.: Finite element approximation of elliptic problems with Dirac measure terms in weighted spaces: applications to one-and three-dimensional coupled problems. SIAM J. Numer. Anal. **50**(1), 194–215 (2012)
18. Scott, R.: Finite element convergence for singular data. Numer. Math. **21**(4), 317–327 (1973)
19. Köppl, T.: Multi-scale modeling of flow and transport processes in arterial networks and tissue. Doctoral dissertation, Technische Universität München (2015)
20. Ojea, I.: Optimal a priori error estimates in weighted Sobolev spaces for the Poisson problem with singular sources. Math. Model. Numer. Anal. **55**, S879–S907 (2021)
21. Apel, T., Benedix, O., Sirch, D., Vexler, B.: A priori mesh grading for an elliptic problem with Dirac right-hand side. SIAM J. Numer. Anal. **49**(3), 992–1005 (2011)
22. Gjerde, I.G., Kumar, K., Nordbotten, J.M., Wohlmuth, B.: Splitting method for elliptic equations with line sources. Math. Model. Numer. Anal. **53**(5), 1715–1739 (2019)
23. Köppl, T., Wohlmuth, B.: Optimal a priori error estimates for an elliptic problem with Dirac right-hand side. SIAM J. Numer. Anal. **52**(4), 1753–1769 (2014)
24. Masri, R., Shen, B., Riviere, B.: Discontinuous Galerkin approximations to elliptic and parabolic problems with a Dirac line source. Math. Model. Numer. Anal. **57**(2), 585–620 (2023)
25. Koch, T., Schneider, M., Helmig, R., Jenny, P.: Modeling tissue perfusion in terms of 1d-3d embedded mixed-dimension coupled problems with distributed sources. J. Comput. Phys. **410**, 109370 (2020)
26. Koch, T., Wu, H., Schneider, M.: Nonlinear mixed-dimension model for embedded tubular networks with application to root water uptake. J. Comput. Phys. **450**, 110823 (2022)
27. Zahedi, S., Tornberg, A.K.: Delta function approximations in level set methods by distance function extension. J. Comput. Phys. **229**(6), 2199–2219 (2010)
28. Köstler, H., Rüde, U.: An accurate multigrid solver for computing singular solutions of elliptic problems. Numer. Linear Algebra Appl. **13**(2–3), 231–249 (2006)
29. Rüde, U., Zenger, C.: On the treatment of singularities in the multigrid method. In: Multigrid Methods II: Proceedings of the 2nd European Conference on Multigrid Methods held at Cologne, October 1–4, 1985 (pp. 261–271). Springer, Berlin (1986)
30. Rüde, U., Köstler, H., Mohr, M.: Accurate multigrid techniques for computing singular solutions of elliptic problems. In: Eleventh Copper Mountain Conference on Multigrid Methods (2003)
31. Gjerde, I.G., Kumar, K., Nordbotten, J.M.: Well modelling by means of coupled 1d-3d flow models. In ECMOR XVI-16th European Conference on the Mathematics of Oil Recovery, vol. 2018, no. 1, pp. 1–12. EAGE Publications BV, Kosterijland (2018)
32. Kuchta, M., Mardal, K.A., Mortensen, M.: Preconditioning trace coupled 3d–1d systems using fractional Laplacian. Numer. Methods Partial Differ. Equ. **35**(1), 375–393 (2019)
33. Trottenberg, U., Oosterlee, C. W., Schuller, A.: Multigrid. Elsevier, Amsterdam (2000)
34. Bader, M.: Space-Filling Curves: An Introduction with Applications in Scientific Computing (vol. 9). Springer, Berlin (2012)
35. Bungartz, H.J., Mehl, M., Weinzierl, T.: A parallel adaptive Cartesian PDE solver using space–filling curves. In: Euro-Par 2006 Parallel Processing: 12th International Euro-Par Conference, Dresden, Germany, August 28–September 1, 2006. Proceedings, vol. 12, pp. 1064–1074. Springer, Berlin (2006)
36. Deuflhard, P., Hohmann, A.: Numerical Analysis: A First Course in Scientific Computation. Walter de Gruyter GmbH, Berlin (2021)
37. Akkurate Behandlung von Singularitäten bei partiellen Differentialgleichungen, H. Köstler, Friedrich-Alexander-Universität Erlangen-Nurnberg, Diplomarbeit.
38. L. Evans, Partial Differential Equations. American Mathematical Society Providence, Rhode Island (1998)

GPSR Compliance

The European Union's (EU) General Product Safety Regulation (GPSR) is a set of rules that requires consumer products to be safe and our obligations to ensure this.

If you have any concerns about our products, you can contact us on ProductSafety@springernature.com

In case Publisher is established outside the EU, the EU authorized representative is:

Springer Nature Customer Service Center GmbH
Europaplatz 3
69115 Heidelberg, Germany

Batch number: 10732090

Printed by Printforce, the Netherlands